Towards the Edge of the Universe
A Review of Modern Cosmology: 2nd Edition

Springer

London
Berlin
Heidelberg
New York
Barcelona
Hong Kong
Milan
Paris
Santa Clara
Singapore
Tokyo

Stuart Clark

Towards the Edge of the Universe

A Review of Modern Cosmology: 2nd edition

Springer

Published in association with
Praxis Publishing
Chichester, UK

Dr Stuart Clark
Director of Public Astronomy Education
University of Hertfordshire
UK

SPRINGER–PRAXIS SERIES IN ASTRONOMY AND ASTROPHYSICS
SERIES EDITOR: John Mason B.Sc., Ph.D.

ISBN 1-85233-098-8 Springer-Verlag Berlin Heidelberg New York

British Library Cataloguing-in-Publication Data
Clark, Stuart (Stuart G.)
  Towards the edge of the universe: a review of modern
  cosmology – 2nd ed. – (Springer–Praxis series in astronomy
  and astrophysics)
  1. Cosmology
  I. Title
  523.1

  ISBN 1-85233-098-8

QB981
.C593
1999

Library of Congress Cataloging-in-Publication Data
Clark, Stuart (Stuart G.)
    Towards the edge of the universe: a review of modern cosmology /
  Stuart Clark. – 2nd ed.
      p.   cm. – (Springer–Praxis series in astronomy and astrophysics)
    "Published in association with Praxis Publishing."
    Includes bibliographical references and index.
    ISBN 1-85233-098-8 (alk. paper)
    1. Cosmology.   I. Title.   II. Series.
  QB981.C593   1999
  523.1–dc21

1852330988

99-35673
CIP

© Praxis Publishing Ltd, Chichester, UK, 1999
Printed by MPG Books Ltd, Bodmin, Cornwall, UK

Cover design: Jim Wilkie
Typesetting: BookEns Ltd, Royston, Herts., UK

Printed on acid-free paper supplied by Precision Publishing Papers Ltd, UK

With thanks to Iain

Table of contents

Colour plates section positioned between pages 142 and 143

Author's preface

When I originally wrote *Towards the Edge of the Universe*, I ended the final chapter by contemplating the changes which might need to be incorporated if the book were to be revised after five years. Little did I know how swiftly cosmology would progress. Now – barely more than two years later – I have completed the second edition.

I will first mention some of the major developments that are discussed within these pages. The results from the Hubble Deep Field – including our first glimpse of star formation rates – have been fascinating, as has the collection of data from the second Hubble Deep Field, in the southern constellation of Tucana. The redshift mapping projects, 2dF and SLOAN, are operational, and both have proved very successful. 2dF has produced a map of 49,636 galaxies, and SLOAN has discovered the most distant quasar yet known. There have been some colossal advances in our understanding of gamma-ray bursts, which were mentioned in the first edition, but which now warrant an entire section. Other fundamental discoveries are suggested by tentative indications that neutrinos do indeed possess a small mass, and that the Universe may not only be expanding but that its expansion is accelerating. This is an amazing prospect and, if proven, may justify the decision of the theoretical cosmologists who refuse to abandon Einstein's cosmological constant.

In the preface to the first edition of this book I stated that the Big Bang theory required further research. That work is still progressing, but the last two years have really been dominated by the observational cosmologists. Infrared and submillimetre telescopes are now the future of observational astronomy. With them we can investigate the deepest realms of the Universe and discover the secrets of the dark ages – the chasm between the galaxies of today and the era of the cosmic microwave background radiation when the first stars and galaxies were forming. Optical work, however, is not defunct. In recent years the Hubble Space Telescope has captured the most exquisite views of the central regions of active galaxies, proving almost beyond doubt that black holes reside in these galactic monsters.

In the first edition of this book I invited readers to respond, and during the past two years I have had the pleasure of receiving a steady flow of comments and ideas. I have also had some interesting conversations, and have (inevitably) been presented with independent theories. If I failed to respond to any of the e-mails or letters, then I offer my apologies. I certainly read them all, and I enjoyed them. With the publication of this second edition, I invite further comments (sclark@star.herts.ac.uk) although I cannot promise to become involved in lengthy and involved correspondence.

I have heard from a number of people who were comforted and inspired by my unorthodox entry into higher education. This little facet of my history was revealed in the first edition's dedication, and since nothing has changed my feelings since I wrote those words, I shall reiterate them.

There are times when the course of life is changed by the influence of an individual person. Only a few years ago I was regularly attending the meetings of Stevenage and District Astronomical Society – at a time when I was a rather frustrated computer programmer. One of the summer meetings featured a lecture on cosmology by Iain Nicolson, and in the sweltering heat of a July evening I learned about the birth and eventual fate of our Universe. As I contemplated – sitting in a room probably only fractionally cooler than the fireball of the Big Bang – I immediately decided to leave my job, turn my hobby into my profession, and study astronomy formally. The snag was that I had only one A-level, which was not even in mathematics or physics (I had failed those) but in computing. However, about a week later I was interviewed at Hatfield Polytechnic (now the University of Hertfordshire), and Iain convinced the Dean of Combined Studies that I was worthy. Under Iain's guidance I obtained a degree and became a member of the university's research team, where I subsequently obtained a doctorate under the guidance of Alan McCall. I continue to research, lecture, write books and promote the public understanding of science, under the generous auspices of the University of Hertfordshire.

I would therefore like to publicly thank my friend and former colleague, Iain Nicolson. I have now written the second edition of a book on the same subject about which Iain lectured on that fateful night so long ago! It seems only right and proper, therefore, to dedicate this work to him.

Stuart Clark
Welwyn Garden City, 7 April 1999

Acknowledgements

All of the following individuals deserve a special mention and my thanks, either for their help in the creation of this book, or for providing background support to my writing and scientific aspirations.

- Clive Horwood, for wanting to publish a second edition.
- John Mason, for his help in defining the contents of this second edition.
- John Watson and David Anderson at Springer-Verlag.
- Bob Marriott, for his copy editing and image processing.
- Nigel Sharp of NOAO, Sue Tritton of ROE, David Malin of AAO and Halton Arp for their generous donation of pictures.
- Caroline Davidson, my agent, for her attention to my career.
- David Axon and Jim Hough, of the University of Hertfordshire, for believing in me and for having the foresight to believe that the public understanding of science is of great importance.
- Steven Young and Chris Courtiour, of *Astronomy Now*, for their continued support of my home internet connection.
- David Hughes and Nial Tanvir, for their reviews of the first edition.
- My friends in the various amateur astronomical societies around the country.
- My colleagues at the University of Hertfordshire, especially Alan McCall and Jim Collett.
- Special thanks to Don Tinkler (my biggest fan), Nick Hewitt, Steve Arnold, Frank Cliff, Mick Hurrell, Christine Dean, Sue Barrett, Christian Kay, Mat Irvine, Jerry Workman and the attendees of my short courses for their loyal support.
- Rush, and other exponents of fine rock music, for keeping me company at the word processor.
- Nik Syzmanek, for introducing me to Dream Theatre.
- Libby – the daughter formerly known as Lizzie – who continues to thrill me with her progress.

Last, but never least, my final vote of thanks goes to Nikki, who has helped in all facets of this book's creation. Through her sustained efforts she has helped to shape the text into a readable exposition. Whatever clarity this text achieves is due in no small part to her commitment and contribution. I am greatly indebted.

List of illustrations, colour plates and tables

1

The astronomical census

Cosmology is the study of the Universe as a whole. It is principally concerned with how the Universe evolved into what it is today and how it will continue to change in the future. Increasingly these days, cosmology is also concerned with how the Universe came into being.

Speculation about the Universe has persisted for the entire length of time that thinking, reasoning human beings have inhabited the planet Earth and looked skywards. The jewel-like pinpricks of coloured light, set in a velvet black expanse, captured peoples' imaginations and inspired them to wonder what they represented. The five 'wandering' stars, which we now know are the planets Mercury to Saturn, added to the mystery of our cosmic abode. The appearance of comets and 'guest' stars, which suddenly flared in intensity before fading back to obscurity, aroused human curiosity. Faced with this wondrous backdrop, human beings could not help but ponder the nature of the heavens.

With this notion in mind, it seems almost unbelievable that the bulk of our modern cosmological knowledge has been gleaned during the twentieth century. Only now is mankind really equipped with both scientific theories and technological instrumentation to undertake the task of understanding the origins of our cosmos. This has led some to claim that we are living in a golden age of cosmology.

In the second decade of the 1900s, Albert Einstein's development of general relativity ended countless thousands of years of speculation, and presented cosmologists with a theoretical tool with which to understand the Universe. And as telescopes continue to become more and more advanced, so they become capable of probing deeper and deeper into space and detecting more and more wonders of the cosmos. The result is that our knowledge of cosmology has risen at almost exponential rates, and at times the flood of ideas and data has been bewildering.

If cosmology is likened to a jigsaw puzzle, then the cosmologists' objective becomes a little easier to visualise. Each object and phenomenon in the Universe is a piece in the puzzle. With each new discovery, so another piece of the cosmic jigsaw becomes apparent. In order to join the pieces of the puzzle together, the cosmologists have a set of rules: the known laws of physics. Unlike the jigsaw player, cosmologists have no picture of the finished puzzle; nor do they even know if they have all the pieces. The nearest they have is an overall impression of what it might look like, assuming that their various theories and hypotheses are correct. Sometimes it becomes obvious that, in order to join together some pieces of the puzzle, more pieces, yet to be discovered, are needed. The required pieces are then shaped according to theory, and sought with telescopes. Occasionally, the searches are successful, and the predicted discoveries are made. At other times, observations do not bear out such predictions, and cosmologists must then re-evaluate their theoretical impression of what the Universe looks like and find new ways to solve the puzzle.

If the basis of cosmology is reduced to a simple 'equation', we obtain:

$$\text{Observations} + \text{Physics} = \text{Understanding of the Universe}$$

In other words, having taken observations, the known laws of physics are used to analyse and understand what has been discovered. This allows astronomers to slot the new observation into its correct cosmological context, and our overall understanding of the Universe increases. In many cases, however, the 'equation' seems more honest if we recast it to be:

$$\text{Observations} + \text{Physics} = \text{Total Confusion}$$

To be fair, reality lies somewhere between the two, and it is this uncertain nature of cosmology which makes it such an exciting field to study.

The currently accepted framework of how the Universe began and subsequently evolved is a collection of interrelated theories known generically as the Big Bang. It is this collection of theories which we will be presenting in this book. However, before we venture on our journey through time and space towards the edge of the Universe, it is worth pausing to take a census of those objects which inhabit the Universe with us. Any cosmological theory must allow for and ultimately lead to the existence of all the known celestial bodies. Starting with a look at our own cosmic backyard, we shall then work outwards.

1.1   THE SOLAR SYSTEM

From times of antiquity, it was known that the heavens contained five wandering stars (the planets), the Sun and the Moon. Uranus, Neptune and Pluto are invisible to the naked eye and were discovered only after the invention of the telescope. In the third century BC the Greek astronomer, Aristarchus, had ventured his opinion that the Earth revolves around the Sun. Five centuries later, however, the work of the Greek astronomer, Ptolemy, who lived in Egypt, suggested a geocentric view of the Universe, in which all of the celestial objects were thought to orbit the Earth. This theory held sway until the sixteenth century. Copernicus finally revived the heliocentric ideas of Aristarchus with the publication of *De revolutionibus orbium coelestium* in 1543. It is important to note here that, in their attempts to understand the Solar System, which to them represented almost the entire Universe, the astronomers of years past were actually practising the science of cosmology.

Figure 1.1. Mars, the fourth planet from the Sun, is a typical rocky world found in the inner Solar System. Mercury, Venus and the Earth all share similar characteristics with Mars. They are all relatively small, dense globes with tenuous atmospheres. (Photograph reproduced courtesy of David Crisp and the WFPC2 Science Team (Jet Propulsion Laboratory/California Institute of Technology) and NASA.)

We now know that the Solar System contains nine planets, numerous moons and an almost countless number of minor bodies, all in orbit around a single star, the Sun. The planets can be divided into two main categories: the rocky terrestrial planets and the gaseous Jovian worlds. Mercury, Venus, the Earth and Mars occupy the first category, whilst Jupiter, Saturn, Uranus and Neptune populate the second. Pluto is something of an oddball and is best classified as being a member of the small icy worlds which make up a large proportion of the Jovian planets' moons.

The terrestrial planets are all relatively small, dense objects (see Figure 1.1). They inhabit the inner regions of the Solar System and have tenuous atmospheres if, indeed, they have any at all. The predominant materials which make up the composition of these planets are silicate compounds and metals, especially iron and nickel. The volatile substances, such as gaseous atmospheres and the water contained in Earth's oceans, were probably deposited by colliding comets after the planets had formed. Volatiles are substances which can change state easily: for example, from gas to liquid, or from solid to liquid. The asteroid belt marks the boundary between the inner and outer Solar System, and represents a planet which was prevented from forming by the gravity of giant Jupiter.

The outer Solar System is the domain of the Jovian planets (see Figure 1.2). These are gigantic worlds, many times larger and more massive than the terrestrial planets. Their compositions are very different and are predominantly composed of light gases and volatiles. This means that, although the planets are very massive compared with the Earth, they are not very dense, since their bulk is spread much more sparsely throughout their oversized volumes. The average density of Saturn is lower than that of water!

Figure 1.2. Jupiter and Saturn are prime examples of gas giant planets. Uranus and Neptune are also gas giants. They populate the outer Solar System and are relatively large (Jupiter is 11 times the diameter of the Earth). They possess large atmospheres, but their overall density is much lower than that of the rocky planets. (Photograph reproduced courtesy of National Astronomical Observatory of Japan.)

The further out into the Solar System one ventures the more prone to volatiles the worlds become. Uranus and Neptune, for instance, are gas giants with very high proportions of the astronomical ices, such as water, methane, and ammonia.

Pluto, a tiny icy world, with a similarly frigid moon, Charon, represents the start of the Kuiper Belt and the Oort Cloud. The Kuiper Belt is a flared disc which is thought to extend into a spherical cloud, the Oort cloud, surrounding the Solar System. Together, they contain vast numbers of icy comets which are thought to be the remains of the material from which the Solar System was formed. Periodically, members of these 'reservoirs' fall inwards towards the Sun and their gases are boiled off by the heat. In general, however, they remain in the furthest reaches of the Solar System.

The motion of the planets was finally described in 1609 by Kepler when he presented his three empirical laws of planetary motion (see Figure 1.3):

> The orbit of a planet about the Sun is an ellipse with the Sun at one focus.
>
> A line joining a planet and the Sun sweeps out equal areas in equal intervals of time.
>
> The squares of the sidereal periods of the planets are proportional to the cubes of their semi major axis.

(In the third law, the sidereal period is the orbital period of a planet, and the semi-major axis is half the longest distance across an elliptical orbit.)

At that time there were no theoretical foundations upon which his laws could be based, yet their validity was beyond doubt because of the way they could be used in order to predict the motions of the planets. A theoretical framework within which

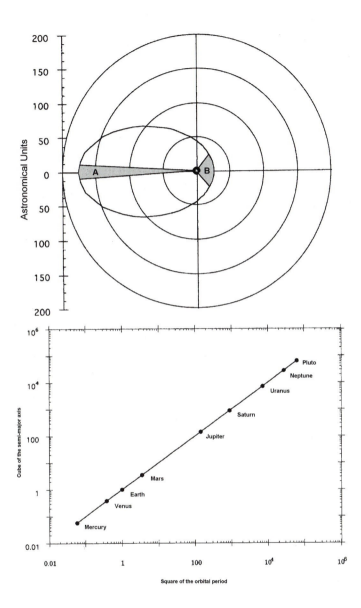

Figure 1.3. Kepler's laws of planetary motion. This triumph of reason by Kepler represented one of the most fundamental advances of his time. The first law states that all orbits are elliptical with the Sun at one focus. The second law states that equal areas are swept out in equal times; the radius vector is an imaginary line joining the Sun and the orbiting body. The third law states that the squares of the sidereal periods are proportional to the cubes of the semi-major axes.

these laws could be hung was finally forthcoming in 1687 when Newton published his three laws of motion:

> A body continues to be at rest or in a state of uniform motion unless acted upon by an external force.

> The change of momentum per second is proportional to and in the same direction as the applied force.

> For every action there is an equal and opposite reaction.

Newton showed how these applied to the planets in the Solar System and how they related to Kepler's laws.

Turning our attention back to the centre of the Solar System and to gain some kind of perspective on it, imagine if everything, except for the Sun, were placed on one side of a scale balance. The Sun is then placed on the other side. Despite grouping all the planets, moons, asteroids and comets together, the Sun would still outweigh them by over 700 times. In fact, the Sun contains 99.8% of the mass contained within the Solar System. So, the matter which constitutes the planets cannot even be thought of as the icing on the Solar System cake!

The Sun itself is overwhelmingly composed of the two lightest gases in the Universe: hydrogen and helium. Therefore we must state that the Solar System, too, is overwhelmingly composed of these two chemical elements. We have to disregard what our eyes tell us from looking around the Earth, and recognise that our entire world, plus everything it contains, is made up of nothing more than the cosmic flotsam produced (as we shall see) as a by-product of stellar evolution.

Our theories of star formation suggest that stars are produced by the gravitational collapse of giant clouds of gas. As the collapse proceeds, the central concentration of matter becomes a star. Other factors, such as the orbital motion of the cloud, mitigate to produce a disc of material around the central concentration. It is in these discs that planets form. Since 1995, there have been a number of extrasolar planets detected. These have ranged from the exotic pulsar planets to the more conventional ones, such as the one orbiting the star 51 Pegasi. In short, planets can be predicted naturally out of star formation theories. Secondly, because the Sun and other stars are far more massive than any planets which may be in orbit around them, it would seem more sensible for cosmologists to switch their attention towards the stars in their attempts to understand the Universe.

1.2   THE SUN AND OTHER STARS

The Sun is a star. It is no different from the pinpricks of light in the night sky except that it is much closer to us. In principle, a star is nothing more exotic than a vast collection of hydrogen and helium gas held together by the force of its own gravity. It contains so much matter, however, that the pressure in its central region is enormous, and the temperature soars to millions of Kelvin. The combination of these factors allows nuclear fusion to take place. This process produces the energy which causes a star to shine and, in the vast majority of stars, the chemical element which is undergoing fusion is hydrogen, the simplest atom which can exist. In stars, hydrogen atoms in the core are fused into helium, the sophomore element of nature.

In most stars, the fusion of hydrogen takes place via the proton–proton chain. The conditions at the core of a star are so intense that atomic nuclei are stripped of their surrounding electrons. In the case of hydrogen, the nucleus is a single proton. The proton–proton chain represents the collisions which must take place between the hydrogen and other nuclei in order to build a helium nucleus.

$$^1_1H + ^1_1H \rightarrow ^2_1H + e^+ + \nu_e$$
$$^2_1H + ^1_1H \rightarrow ^3_2He + \gamma$$
$$^3_2He + ^3_2He \rightarrow ^4_2He + ^1_1H + ^1_1H \tag{1.1}$$

In step one, two hydrogen nuclei (protons) collide to form a heavy hydrogen nucleus (known as deuterium), a positron and a neutrino. In step two, another proton collides with the deuterium nucleus to form a light helium nucleus (called tritium) and a gamma-ray. The final step of the reaction involves the collision of two tritiums, which forms a helium nucleus and releases two protons.

A few, more unlikely routes to form helium are also possible, but in a star such as the Sun the above reaction route is dominant. The energy emitted by these reactions is produced because the reactions convert a small amount of mass into energy. In all cases the summed mass of the reactants (those particles on the left-hand side of the arrows) is less than the summed mass of the products (those particles on the right-hand side of the arrows). This difference in mass is known as the mass deficit, Δm, and is turned into energy, E, in accordance with Einstein's famous equation,

$$E = \Delta mc^2 \tag{1.2}$$

The constant of proportionality in this equation is the square of the speed of light.

Study has shown that there are numerous different types of star strewn through space. An excellent way in which to classify them initially is by their colour, which instantly reveals the star's surface temperature. For example: yellow stars such as the Sun have surface temperatures of 6,000 K; cool, red stars have surface temperatures of about 4,000 K; and the hottest stars in the Universe are blue, with surface temperatures in excess of 20,000 K.

The precise classification depends upon other characteristics such as luminosity, radius and a complete analysis of the star's spectrum. These properties help determine the mass of the star in question and the fusion reactions taking place in its core. To help in the classification a diagram, known as the Hertzsprung–Russell diagram after its devisers, has been developed (see Figure 1.4). Upon this diagram every known star has a place which is determined by its brightness and surface temperature. As a star evolves, the nuclear reactions taking place within it change in accordance with which elements are available for fusion, and the star changes its classification. The changes can be charted on the H–R diagram, and produce evolutionary tracks which represent the different classifications a star occupies during its lifetime. Just like an Earth plant or creature, a star has a life cycle; it is

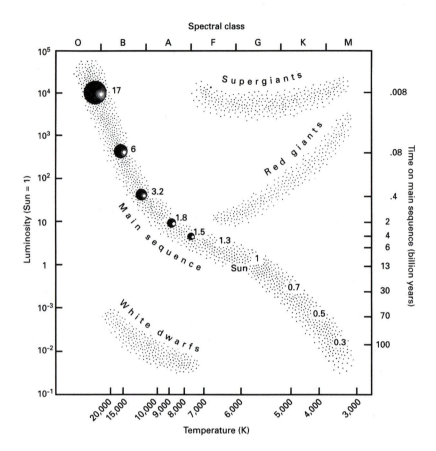

Figure 1.4. Every star in the Universe has a place on the Hertzsprung–Russell (H–R) diagram. The sweeping band of stars running from top left to bottom right across the centre of the plot is the main sequence, where the majority of stars can be found. The main sequence represents those stars, such as the Sun, which are in the stable, hydrogen-fusing stage of their existence. The numbers along the main sequence are stellar masses, in units of the Sun's mass.

initially formed and eventually dies. During the course of its life, a star manufactures progressively heavier chemical elements because of the nuclear reactions taking place in its core.

The eventual fate of a star is virtually sealed by the mass it contains when it is formed. A star which contains less than five times the mass of the Sun will fuse hydrogen into helium for many millions – or in some cases billions – of years. When its supply of hydrogen runs out, internal conditions will change; the star will become a red giant star, and will begin fusing helium into carbon. As the helium runs out, so the star will die. It will return its outer layers of hydrogen and helium into space, causing something that is deceptively called a planetary nebula. What is left is a small, compact object known as a white dwarf. Stars of mass greater than five times that of the Sun are individually more significant on the cosmological scale. They are present in the Universe in far fewer numbers, and have much shorter lives, than their less massive cousins. Whereas smaller stars cannot fuse the carbon produced by helium fusion, these larger mass stars can. In fact, they can fuse a great many chemical elements through a complex pattern of interactions which are made possible by the extreme conditions in their interiors. Iron, however, is a chemical element which marks a watershed in nature. All elements lighter than iron can be fused in order to emit energy, but from iron upwards energy needs to be supplied in order for the elements to be fused.

In the hearts of massive stars, iron builds up like ash in a fire. This is because nowhere near enough energy necessary to fuse iron is available in a star's core. When the amount of iron reaches about 1.5 times the mass of the Sun, its atomic structure breaks down and causes the star to collapse. Material from the star's outer layers falls downwards towards the shrunken core and impacts its surface. This is followed by a huge explosion, which blows the star to pieces and supplies all the excess energy required to fuse all the elements heavier than iron. The result is a supernova (see Figure 1.5, colour section). If anything remains of the original star, it may be a fragment of its collapsed core – a neutron star. Alternatively, the collapsed core may have been so large that it becomes a black hole.

When they explode, these massive stars seed space with heavier elements. In short, they change the composition of the Universe. We know from measurement that the composition of the Universe today is 75% hydrogen, 23% helium and 2% (at most) everything else. This 2% consists of all the atoms apart from the hydrogen and helium, which make us, the Earth and the Solar System.

So the composition of the Universe has changed with time due to the action of stars. An immediate consequence of this is that stars in the early Universe could not possibly have had planets such as the Earth in orbit around them because the Universe did not contain sufficient heavier elements from which to make rocky worlds. Planets, when they did form, would have been gaseous bodies, similar in outward appearance to Jupiter.

The distribution of stars within the Universe was a major concern to cosmologists of the 1700s and 1800s. The issue was so contentious that it was only finally decided in the second decade of the twentieth century, due to the work of Harlow Shapley.

1.3   THE GALAXY

Originally, it was believed that stars were positioned randomly throughout space. With the advent of the telescope, however, it became obvious that this was not the case. The Milky Way, known throughout history as being a misty band of light which stretched across the sky, was shown by Galileo's telescope to be composed of hundreds upon hundreds of faint stars. It seemed obvious to him when he peered at them, that stars were grouped together in some way or another.

The suggestion that the Sun was part of a much larger, disc-shaped grouping of stars was first made in 1750 by Thomas Wright in his published work, *Theory of the Universe*. William Herschel attempted to map the distribution of stars in the 1780s and found differences of up to a factor of 600 in the stellar densities of different areas of the sky.

The breakthrough in our understanding of the grouping of stars in the Galaxy and especially the Sun's place within it, came at the beginning of the twentieth century, when Harlow Shapley studied the distribution of distant spherical collections of stars, called globular clusters. He discovered that most lay in the direction of the southern Milky Way. If Shapley's globular clusters were to be enclosed by a gigantic sphere, then the centre of that sphere would not be centred on the Sun. Instead, it would be some 25,000–30,000 light years in the direction of Sagittarius. Shapley made a bold assumption and concluded that this marked the centre of the Galaxy, since the globular clusters must in orbit around it. This conclusion was indeed correct, and changed our view of the Universe from being heliocentric to galactocentric.

In the mid-twentieth century, serious attempts were made to map the structure of our Galaxy. Many lines of investigation have shown that the disc contains spiral 'arms' of stars which swirl around the centre of our Galaxy like an octopus on a turntable (see Figure 1.6)! More recent work on the rates at which stars orbit the Galactic centre has thrown up some startling conclusions about the nature of matter contained within the Galaxy.

It appears that the stars are being gravitationally pulled by a vast quantity of matter in a spherical halo surrounding the disc and nucleus which traditionally constitute the Milky Way. It has proved impossible to detect this matter by any other means except by the gravitational influence it would appear to exert on the stars in the Galaxy's disc. Hence, its exact nature is somewhat controversial at present, and many theories abound. What seems certain, however, is that the visible matter in the Galaxy – such as that which is contained in the stars – is vastly outweighed by the invisible material. This inequality may be as much as 99:1, and so we can no longer think of our Galaxy as being just a collection of stars. Instead, our modern cosmological view of the Universe pictures the Galaxy as being a region of the Universe which has a denser-than-average collection of matter within it. If we extend this line of reasoning, stars are then nothing more exotic than denser-than-average regions of the Galaxy. Exactly how these Galactic regions became so dense is one of the driving questions in cosmology.

1.4   THE LOCAL GROUP AND OTHER GALAXIES

At about the same time as Harlow Shapley was carrying out his research on the Milky Way, another debate was raging. The use of telescopes over two centuries had revealed a considerable number of celestial objects which were not stars. They became known as 'nebulae', and the brightest of these non-stellar objects had already been catalogued by Charles Messier in 1781. Many are still known today by their Messier numbers – such as M42, the Orion nebula.

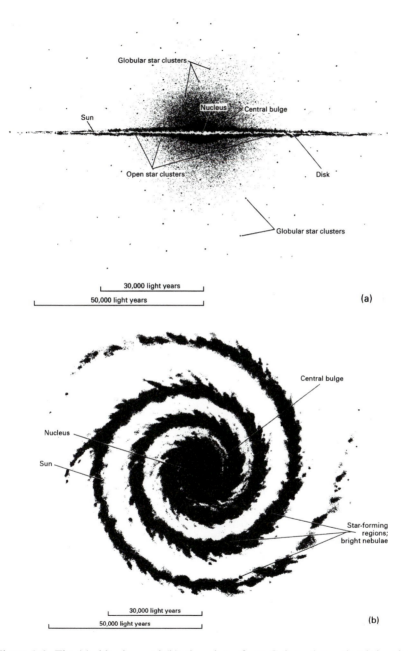

Figure 1.6. The (a) side view and (b) plan view of our Galaxy shows that it is spiral in form and that our position is offset from the centre. The Sun is located near the inner edge of the Orion spiral arm. The sweeping band of stars in the night sky – the Milky Way – is our view through the plane of the starry disc. The centre of the Galaxy is located in the direction of the constellation Sagittarius. (Adapted from Ferris, T., *Galaxies*, Stewart, Tabori and Chang, 1982.)

In the early part of the twentieth century, observers divided the Messier objects into two distinct categories. Some were obviously clouds of gas, and these seemed to congregate near the Milky Way. Others showed an elliptical or spiral shape and seemed to occur at random throughout the sky. This latter class of object became known as spiral nebulae, and some astronomers suspected that they were collections of stars, just like the Milky Way, but viewed from a much greater distance. Controversy raged between these astronomers and those others who thought they were gas formations or hazy stars within the Milky Way. Interestingly enough, Harlow Shapley – who had shown such remarkable insight concerning our own Galaxy – was vehemently (and incorrectly) opposed to the point of view which stated that spiral nebulae were galaxies in their own right!

As far back as 1755, philosopher Immanuel Kant had expressed the view that these distant objects were perfect analogies to the Milky Way. Yet the matter was only finally ended in 1924 when Edwin Hubble showed that some of the nearby spiral nebulae contained variable stars, similar to those within the Milky Way but much further away. Hubble went on to classify galaxies according to their appearance. He found three broad types: spirals, barred spirals and ellipticals. Any galaxy which did not fit into this scheme was termed irregular. Hubble meticulously mapped the sky, looking for faint galaxies in order to understand how they were distributed throughout space. Statistical analysis of his maps showed him that galaxies tend to clump together into groups and larger clusters.

The Milky Way is part of a small group of galaxies known as the Local Group (see Figure 1.7). It consists of approximately 30 known members. The hierarchy does not stop at groups and clusters, however. Clyde Tombaugh, the discoverer of Pluto, recognised that even clusters themselves tend to cluster together! They form massive collections known as superclusters (see Figure 1.8) which, as has been shown in the last few decades, stretch through space in chains and filaments. This represents such a fundamental departure from the random distribution of galaxies, that it provides one of the most stringent constraints which must be placed upon any modern cosmological theory if it is to be successful in explaining the Universe as a whole. Another important problem for cosmology is whether the individual galaxies, or the clusters which contain them, formed first.

Modern cosmology studies the distribution of matter on scales which are far larger than individual galaxies. This means that for many purposes we can imagine the galaxies as being the most fundamental objects in the Universe. In fact, cosmology often assumes that the matter contained in the Universe can be thought of as a perfect gas. There are so many different types of individual galaxies, however, that the cosmological question of how they became the way they are is still important.

1.5   ACTIVE GALAXIES AND QUASARS

The first half of the twentieth century witnessed a greatly increased interest in galaxies, and they were studied in great detail. During this time, astronomers began

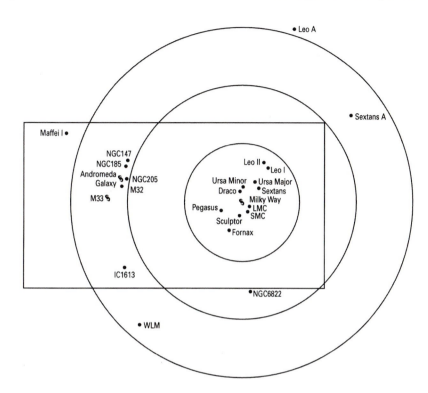

Figure 1.7. The Local Group is dominated by two spiral galaxies: our own Milky Way and the Andromeda galaxy (M31). Each of these galaxies is attended by smaller satellite galaxies and a number of dwarf galaxies. M33 is another spiral galaxy, and a recent discovery is the barred spiral galaxy Dwingaloo 1. (Adapted from Ferris, T., *Galaxies*, Stewart, Tabori and Chang, 1982)

to notice that, in terms of their radiation output, some galaxies were distinctly different from others. In 1943, for instance, astronomer Carl Seyfert noted a number of spiral and barred spiral galaxies which appeared to have nuclei much brighter than normal. These became known as Seyfert galaxies, and were the first of the active galaxies to be discovered. (The term 'active' is used to describe a galaxy which is producing very much more radiation than that produced by its stars.)

The following year, in 1944, amateur astronomer Grote Reber was observing the heavens with a home-made radio telescope when he detected an incredibly powerful radio source emanating from something in the constellation of Cygnus. This was the first recognition of the radio galaxies which, as the name implies, radiate the majority of their energy output at radio wavelengths. In the 1950s, the position of Cygnus A was gradually refined until it was finally identified with a fairly dim elliptical galaxy. Unlike the Seyfert galaxies, the radiation from radio galaxies does not come directly from the nucleus. Rather, it is released from gigantic 'lobes' which exist on either side of the host galaxy. They appear to be fed by a pair of high-energy jets being ejected in opposite directions from the active galactic centre.

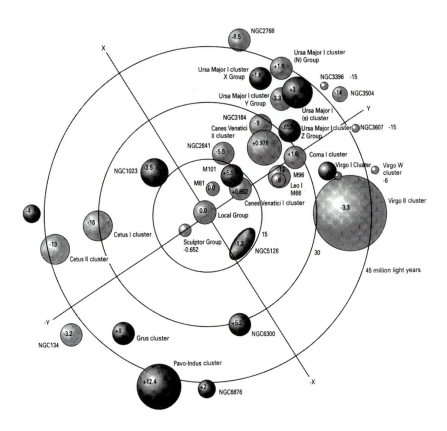

Figure 1.8. Schematic of the Local Supercluster. The Local Group is totally dominated by the other, much larger clusters of galaxies. Of these, the Virgo cluster is the most conspicuous from Earth, because its members can be readily seen in a modest telescope set in the direction of the constellations Virgo and Coma Berenices. The figure accompanying each cluster is its distance in light years above or below the X–Y plane. (Adapted from Ferris, T., *Galaxies*, Stewart, Tabori and Chang, 1982.)

In 1963 a third class of active galaxy was identified by a team of radio astronomers led by Cyril Hazard. These peculiar celestial objects – quasars – superficially appear to be stars but, under analysis, reveal that they are extremely distant and incredibly luminous galactic objects (see Figure 1.9). They share many of the same features as Seyfert and radio galaxies, but emit even more radiation. Curiously, quasars seem to be much more prevalent at greater distances, which has led to the belief that they are of cosmological importance. The fact that they once existed in great numbers, but no longer do so, may be of crucial importance to our understanding of how galaxies came into being and evolved with time.

On average, one in every ten known galaxies is active; that is, the majority of its radiation output is not being produced by stars. Instead, some other physical

Figure 1.9. On ordinary photographs of the sky, quasars, such as the one identified here by the two vertical bars, cannot be distinguished from stars in the same field. Only after more careful analysis (examination of their spectra) is it revealed that quasars are extremely distant and incredibly luminous galactic objects. (Photograph reproduced courtesy of Mount Palomar Observatory.)

mechanism is releasing copious amounts of radiation, up to 1,000 times greater than that emitted by entire galaxies, and often from a region with approximately the same diameter as our own Solar System!

There are many different ways of classifying active galaxies, apart from the three types which have already been mentioned. The crucial question is: were all galaxies once active? Perhaps all of the so-called 'normal' galaxies – our own Milky Way included – once passed through an active phase. Now, however, the active core has fallen dormant and the galaxies' emissions are dominated by starlight. If this is the case, then it would seem that all young galaxies pass through an active stage because of the distribution of quasars throughout space. If, on the other hand, active galaxies are intrinsically different from normal galaxies, then this conclusion too would provide a valuable insight into the early Universe. Why, for example, would the Universe promote the evolution of quasars billions of years ago, but not now?

Active galaxies are fascinating objects to study because the answers to some intensely perplexing cosmological puzzles seem wrapped up in their mysterious cores.

1.6   THE SPACETIME CONTINUUM

In thinking about the Universe, modern cosmology demands that we consider not only the objects it contains. As well as thinking about the radiation content of the Universe (discussed in the next chapter) we must also consider space itself. Kepler's laws of planetary motion were the first to describe how objects move through space. They were kinematic in nature because they made no attempt to explain why the planets moved. Newton went one stage further by considering the force which was causing the motion: gravity.

Newton's laws of motion led him to theoretically derive his universal law of gravitation, which gives the magnitude of the gravitational force, F, between two masses, m_1 and m_2, which are separated by a distance, r:

$$F = G \frac{m_1 m_2}{r^2} \tag{1.3}$$

where G is the universal constant of gravitation, $6.67 \times 10^{-11} Nm^2/kg^2$. This work of Newton's not only ushered in the era of theoretical science but also proved that planetary motion was caused by the familiar terrestrial force of gravity. It was a triumph of unification, and (as we shall see in the next two chapters) the task of unifying nature's diverse phenomena continues to this day. Newton used his equations to deduce that orbits could be circles, ellipses, parabolae or hyperbolae. These are known as conic sections, because their shapes can all be obtained by slicing through a cone (see Figure 1.10).

Another great unification took place early in this century between space and time. This was made possible by the concept of spacetime, developed by Hermann Minkowski. Instead of thinking of space as being three-dimensional, with time as a separate property, spacetime postulates a four-dimensional framework in which the time ordinate can be assigned the same units as distance by multiplying it by the speed of light. In 1915, Einstein built upon spacetime and presented the general theory of relativity. This was also an extension of Einstein's previous work – the

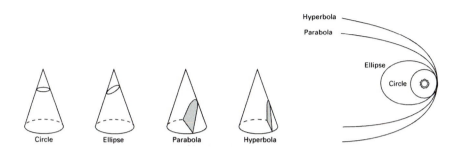

Figure 1.10. All orbits are represented by one of the conic sections. Different conic sections are obtained by slicing a cone at different angles. Closed orbits are circles and ellipses, obtained by slicing the cone at zero or shallow angles to the axis of symmetry. Open orbits are the parabolae and hyperbolae, obtained by slicing the cone at steep angles to the axis of symmetry. (Adapted from Kaufmann, W.J., *Universe*, W.H. Freeman, 1987.)

special theory of relativity – and it helped to explain gravity. It finally gave astronomers a tool with which to study the Universe on its largest scale.

In 1929 a very important conclusion was reached, based upon a set of interesting observations. Hubble's work on galaxies yielded the result that all the galaxies, apart from those in the Local Group, were receding from the Milky Way. The further away the galaxy, the faster it was receding.

The receding motion of the galaxies was described by Hubble's equation, which related the recessional velocity of a galaxy, v, to its distance, d:

$$v = Hd \tag{1.4}$$

H is known as the Hubble constant and has a value which is thought to lie somewhere in the range 50–100 km/s/megaparsec.

The general theory of relativity could be made to agree with this conclusion if one major caveat were to be placed upon the explanation. Although it appeared as if it were the galaxies which were moving through space, in actuality it was the space between the galaxies which was expanding. So the galaxies were being driven apart in the same way that raisins in a dough mixture are moved away from one another when the bread rises (see Figure 1.11). Interestingly enough, when Einstein had derived the equations which define the spacetime continuum, the equations did not allow the Universe to be static; it had to be either expanding or contracting. Since, at the time, Hubble had yet to prove the expansion of the Universe, Einstein had recast his equations for presentation so that they included a new constant which would hold the Universe static. When Hubble presented his results, Einstein immediately removed his cosmological constant and referred to it as the greatest mistake of his life!

General relativity provides us with a means of visualising how gravity is created, by explaining the way mass interacts with spacetime. The theory states that massive

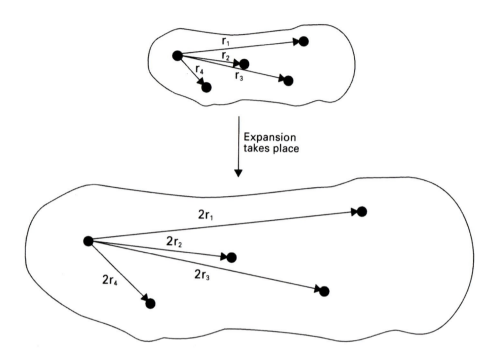

Figure 1.11. The Universe is thought to be in a state of uniform expansion. This is similar to a rising dough mixture which contains raisins. Because the dough between the raisins expands, the further the raisins are from one another, the faster they will move apart.

objects distort the spacetime continuum into curves in much the same way as a heavy ball might deform a rubber table-top (see Figure 1.12). Anything which approaches too close rolls down the curve as if it were being attracted to the massive object. Thus, there exists an intimate relationship between space and matter. The curvature of the spacetime continuum caused by mass in turn creates gravity, which dictates how other masses move through it.

One of the universal paradoxes which cosmologists have had to deal with is that, although gravity is by far the weakest of the four fundamental forces of nature, it dominates the Universe on its largest scales. The strong nuclear force and the weak nuclear force operate only over the distance of an atomic nucleus, whilst the electromagnetic force cancels out over large distances. With nothing left in competition, gravity can sculpt the Universe on all but the smallest of scales. Indeed, it is gravitational forces which are responsible for every structure we have so far discussed in this chapter. As far as the future of the cosmos is concerned, gravity has also sealed the fate of the Universe. All that remains is for the cosmologists to exactly determine the nature of that fate!

Cosmology is a rich field of study which encompasses all scales of the Universe. The current theory of the origin of the Universe – the Big Bang – is understandable

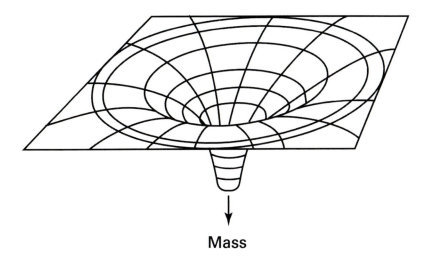

Mass

Figure 1.12. The spacetime continuum can be visualised as a rubber sheet onto which are placed heavy objects. These objects (stars, galaxies and so on) deform the rubber sheet, causing it to curve. Any object which strays too close, rolls down the curve and touches the object causing the curvature. This is the general relativistic analogy to gravity. (Adapted from Berry, M., *Principles of Cosmology and Gravitation*, Adam Hilger, 1989.)

only by thinking in terms of quantum theory: the method we have for understanding the Universe on its smallest scale. The long-term future of the Universe is understandable only if we use general relativity: a theory for understanding the Universe on its largest scale. Cosmology is, literally, universal in content and scope. It is the science from which all others are born, and we gain access to it by taking our minds on a journey from here towards the edge of the Universe.

2

The tools of the trade

The astronomical census presented in Chapter 1 has been made possible by the collection of electromagnetic radiation from space. Astronomy, by its very nature, is an observational science, not an experimental one. As much as some of us would like to, astronomers are not free to journey to these far distant objects in the exotic depths of the Universe and set up experiments to learn their secrets. Instead, we must rely on collecting the radiation which has been released by these celestial bodies.

The detection of electromagnetic radiation is by far the most advanced of our methods for examining the Universe. Since the dawn of human existence, mankind has done this quite naturally, because our eyes are really rather sophisticated detectors of visible electromagnetic radiation. Technologically, the study of visible light began in 1609 when Galileo used a telescope to look at the night sky. Today, a plethora of collecting and detecting devices is capable of sampling the radiation from many areas of the electromagnetic spectrum.

2.1 ELECTROMAGNETIC RADIATION

In the same way that the very fact of the Universe's existence has caused many to wonder about it, so the nature of light has also puzzled mankind throughout history. Plato, Aristotle and Pythagoras all wondered about it, but nothing really came of their pontifications. Between the 1300s and 1600s, European researchers concentrated on the development of lenses and mirrors without really wondering what constituted the light they were studying. Whilst doing this, however, many became aware of the way in which rays of light moved through the air and interacted with other rays of light. Gradually, curiosity was raised and the physical nature of light was considered.

In 1665, whilst attempting to understand a series of observations which generated the phenomenon of diffraction, Robert Hooke proposed that light is a rapid vibration of the medium through which it is passing. Hooke's presentation marks the beginning of the wave theory of light which is still in use today.

Many excellent contributions were made to the burgeoning science of optics, but

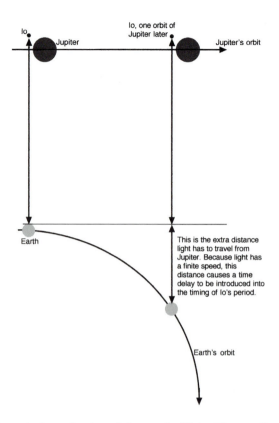

Figure 2.1. Rømer's determination of the speed of light. The peculiar differences which Rømer measured in Io's period led him to conclude that light must travel with a finite velocity.

possibly the next two greatest milestones were the proof that light travelled with a finite velocity and that it was a type of electromagnetic wave. Both conclusions were finally and irrefutably reached in the middle of the nineteenth century.

From the days of Galileo, scientists had been trying to measure the speed of light. Those early attempts now seem crude, and their failure to show any time delay in light's propagation from one location to another led some to believe that it travelled infinitely fast. In 1675, however, Danish astronomer Ole Christensen Rømer observed an astronomical phenomenon which he could explain only if the speed of light were finite. He was trying to measure the orbital period of Jupiter's moon Io, but kept obtaining different results. In the time taken for Io to orbit Jupiter, the Earth was moving in its own orbit around the Sun. The distance between the Earth and Io was therefore not constant. If light travelled instantaneously then this would make no difference; but if not, the orbital period would be affected because of the additional time taken for light to cross the varying distance between Io and Earth (see Figure 2.1). It was the first direct evidence that light did not propagate instantaneously, and it even yielded a crude estimate of its velocity. Just over 50 years later, another phenomenon – aberration – was discovered by the Englishman James Bradley. This, too, was explainable only in terms of light travelling with a finite speed. Like Rømer before him, Bradley also calculated a speed for light.

The first relatively accurate measurement of the speed of light was made by Frenchman Armand Hippolyte Louis Fizeau in the suburbs of Paris during 1849. His calculated figure was 315,000 km/s, which, compared with today's accepted figure of 299,792 km/s, was good for the technology of the day.

At about the same time as these measurements in optics were being undertaken, another seemingly unrelated field of physical study was also progressing rapidly. Michael Faraday had applied his great mind to the study of electricity and magnetism. Within Faraday's lifetime, Hans Christian Oersted had established a link between electricity and magnetism when, in 1820, he noticed that an electric current in a wire affects a magnetic compass needle. Faraday himself noticed a link between light and electromagnetism when he discovered that a strong magnetic field could alter the properties of a ray of light.

Building upon this information and the collected results of many other experimentalists, another physicist, James Clerk Maxwell, expertly extended the work and eventually distilled the behaviour of electromagnetism into a series of four theoretical equations. His work stands today as one of the greatest examples of theoretical physics – second only to Einstein's general theory of relativity. In the course of his investigations, Maxwell became aware that his equations predicted that electromagnetism could theoretically travel in the form of transverse waves. Solving his equations to produce the speed of the wave, Maxwell discovered that it travelled at Fizeau's measured speed of light! This was too big a coincidence, and Maxwell felt compelled to draw the conclusion that light was an electromagnetic disturbance propagated by a wave (see Figure 2.2). Not only that, but light was concentrated into such a restricted set of wavelengths that other types of electromagnetic radiation

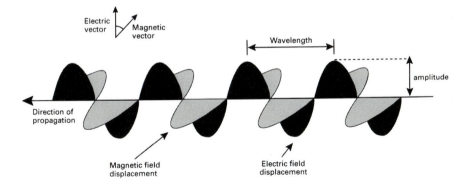

Figure 2.2. An electromagnetic wave consists of two disturbances which propagate through space. The electric vector is an oscillating disturbance at right angles to a similarly oscillating magnetic disturbance. Together they travel through space at the speed of light.

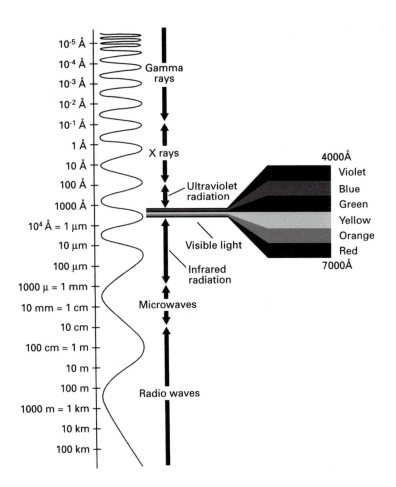

Figure 2.3. The electromagnetic spectrum. All types of electromagnetic wave are similar to one another. They are distinguished only by their wavelength, which dictates how the wave will interact with matter. Artificial divisions define these differences, although in reality the transition from one type of electromagnetic wave to another is smoothly continuous. 1 nanometre (nm) = 10 Ångstroms (Å). (Adapted from Kaufmann, W.J., *Universe*, W.H. Freeman, 1987.)

(with longer and shorter wavelengths) could be predicted to exist. In 1888, experimentalist Heinrich Rudolf Hertz dramatically proved Maxwell's theory of electromagnetic radiation by producing the first radio waves.

Electromagnetic radiation is now artificially divided into seven categories: radio waves, microwaves, infrared, visible, ultraviolet, X-rays and gamma-rays. Fundamentally, there is no difference between these types of radiation; they simply have different wavelengths and frequencies (see Figure 2.3).

The defining properties of a wave – the wavelength, λ, and the frequency, f – are interlinked by the speed of the wave's propagation, c, in the following way:

$$c = f\lambda \qquad\qquad\qquad\qquad\qquad\qquad\qquad\qquad (2.1)$$

In a vacuum, the speed is approximately 3×10^8 m/s.

2.2   ELECTROMAGNETIC SPECTRA

The entire gamut of electromagnetic radiation is known today as the electromagnetic spectrum. The word 'spectrum' was first coined by Isaac Newton who was the first person to produce and study the optical spectrum. The word itself is Latin for 'ghost' or 'apparition', and was used by Newton to describe the ephemeral pattern of colours which danced on the wall of his darkened room at Cambridge when he placed a prism in the path of a light-beam.

Initially, the spectra observed were simply continuous bands of colour. In 1802 W.H. Wollaston observed a few dark lines in the solar spectrum and assumed that they were gaps between the colours, but in 1814 Joseph Fraunhofer magnified a spectrum of sunlight and discovered numerous dark lines. He investigated the nature of these lines and catalogued them. He is therefore widely credited as making the first observation of an absorption spectrum. Later that century, emission spectra were observed by two chemists, Robert Bunsen and Gustav Kirchhoff. They passed the light from chemical flame tests through a prism and discovered that the spectra were simple patterns of coloured lines rather than dark lines superimposed upon continuous colours. Further investigations showed that each element had a characteristic pattern of these emission lines. It was also realised that every emission spectrum had a corresponding absorption spectrum (see Figure 2.4).

It appeared that, in some way, atoms could only absorb and emit at specific wavelengths of radiation. By now, physicists also recognised that a truly continuous spectrum was produced only by heating an object and letting it radiate that heat as electromagnetic radiation. Thus, this method of producing electromagnetic radiation became known as thermal emission. Thinking of light as a wave, however, made it impossible to understand why atoms and radiation should behave in this way. The production of thermal radiation is characterised by three empirical laws which assume that the object being heated is a perfect absorber (and perfect emitter) of radiation. Since a perfect absorber would not reflect any radiation, its colour would be black. Thus, an object of this type has become known as a black body. All radiation absorbed by the object will be converted into heat energy and then re-radiated at lower frequencies. As well as a solid object at a specific temperature, a dense ideal gas can also radiate as a black body if its constituent atoms and molecules are in thermal equilibrium. In most astronomical and cosmological cases, it is emission by a dense ideal gas in thermal equilibrium which produces black-body radiation. For example, stars are well approximated by black bodies and, as we shall see, the cosmic microwave background radiation is also characterised by a black-body curve. Hence, from now on, this book will assume that Planck (black-body) curves are being produced by dense ideal gases in thermal equilibrium.

The first law which describes the process of black-body thermal emission is known

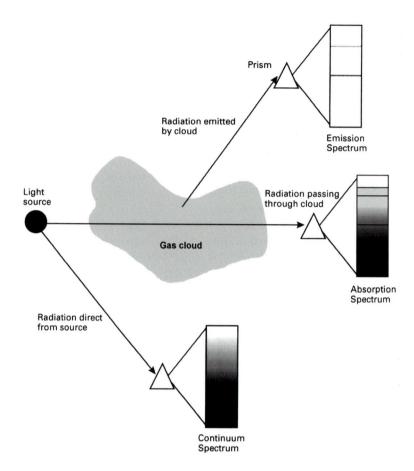

Figure 2.4. Emission, absorption and continuous spectra. The continuous spectrum is produced by a hot, solid body or by a hot, dense gas. Emission spectra are produced by a hot, tenuous gas, and absorption spectra are produced by cold clouds of tenuous gas, illuminated from behind.

as the Stefan–Boltzmann law. It determines the total radiated power per unit area, R, for a black body at a given temperature, T:

$$R = \sigma T^4 \tag{2.2}$$

The proportionality is maintained by the Stefan–Boltzmann constant, defined as $\sigma = 5.67 \times 10^{-8} \ \text{W/m}^2\text{K}^4$.

The second law is known as the spectral radiancy, and describes how the intensity of radiation changes with wavelength and temperature. Graphs constructed for the spectral radiancy at specific temperatures produce Planck or black-body curves (see Figure 2.5). When the spectral radiancy is integrated over all wavelengths it equates with the Stefan–Boltzmann law.

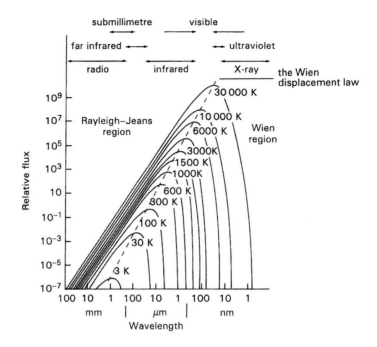

Figure 2.5. Planck curves and the Wien displacement law. Any hot body emits radiation in accordance with these Planck curves. The hotter the body, the more radiation is emitted and the shorter the wavelength of peak emission. This change in the wavelength of peak emission is described by the Wien law.

The third and final, empirical law is the Wien displacement law. It states that the wavelength at which the maximum intensity of radiation is released, λ_{max}, is inversely proportional to the temperature of the radiating body:

$$\lambda_{max} = \frac{W}{T} \qquad (2.3)$$

The Wien constant, w, is equal to 2,898 µm K.

At the turn of the twentieth century, physics was facing the challenge of finding a formula which could fit the Planck curves and be totally derivable from a theoretical basis. The best attempt at fitting the curve that classical physics could offer was the attempt made by Lord Rayleigh and later modified by James Jeans. It could fit the curves only at very long wavelengths. A second attempt, by Wilhelm Wien, fitted better at shorter wavelengths but failed to fit at long wavelengths. Wien departed somewhat from classical theory, however, by assuming that there was an analogy between the spectral radiancy curves and the Maxwell speed distribution curves for the molecules of an ideal gas.

Max Planck solved the problem by interpolating between the two models and presenting a formula which fitted the radiation at all wavelengths. He then set out to derive his equation again, this time from a simple set of theoretical assumptions. He succeeded brilliantly, and presented his theoretical equation which perfectly matched the spectral radiancy curves:

$$I(\lambda,T) = \frac{2\pi c^2 h}{\lambda^5} \frac{1}{e^{hc/\lambda kT} - 1} \tag{2.4}$$

Planck's equation used two new constants: the Boltzmann constant, $k = 1.38 \times 10^{-23}$ J/K; and the Planck constant, $h = 6.63 \times 10^{-34}$ Js. In order to produce his equation, he had to make a radical assumption about the way in which atoms release radiation. Classically, it had been assumed that atoms could emit photons of any energy. Planck introduced the idea that atoms could release radiation at only certain predefined energies. Einstein proposed that light could sometimes be thought of as particles (photons) and, conversely, Erwin Schrödinger proposed that matter could sometimes be thought of as waves. Using these ideas – which became generically known as quantum theory – Niels Bohr explained the possible positions an electron could assume around a hydrogen nucleus by consideration of their emission spectra. The photon concept allows the energy carried by each photon, e, to be calculated in terms of the light's frequency, f:

$$e = hf \tag{2.5}$$

The key to quantum theory is wave–particle duality. It states that light – traditionally thought of as a wave – can sometimes be a particle, and that electrons – traditionally thought of as particles – can sometimes behave as waves. As physicists and astronomers, we are free to regard light and electrons as particles or waves, in order to solve the problem at hand. This seems to be quite a fundamental liberty to take on the part of the scientists, and the question naturally arises: what is the true nature of an electron? Is it a wave or a particle? The answer is that, just like light, it is both and yet neither.

Quantum theory states that everything has both a particle nature and a wave nature. This book is not only a solid object, but is also a wave formation. The key to why it behaves like a solid object, however, is that it is very much more massive than an electron. Electrons contain such tiny masses that when the quantum theory equations for their behaviour are solved, neither wave nature nor particle nature shows dominance. Hence, they display the qualities of both. If the quantum equations were solved for this book, however, its solid, particle-like nature would overwhelm the wave nature.

The concept of quantisation and wave–particle duality has shown that electrons can exist only around atomic nuclei in certain states defined by the quantum nature of the atom. When radiation comes into contact with electrons around atoms, the photon will be absorbed only if it contains enough energy to allow the electron to jump to a new energy state. Later, when the electron jumps back to its original level, the energy is re-emitted as a photon with a specific wavelength. Thus, the process responsible for the production of absorption and emission spectra is finally understandable, and the reason why there are line spectra follows naturally from the explanation.

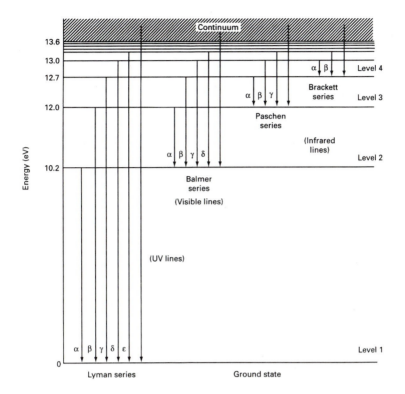

Figure 2.6. Energy levels around a hydrogen nucleus. In the simple Bohr picture of the atom, energy levels are said to be quantised; that is, the electrons may take up only certain specific energies. For the hydrogen atom, an energy-level diagram can be drawn showing all available energy levels and the series of spectral lines (Lyman, Balmer, Paschen and Brackett) produced by transitions between them. Electrons jumping from higher energy levels to lower energy levels produce emission lines, while those jumping from lower-energy levels to higher-energy levels produce absorption lines.

The production of line spectra occurs because the electrons are jumping between quantised energy levels, but at no stage are the electrons losing their association with their parent nuclei. Thus, these transitions are known as bound–bound transitions (see Figure 2.6).

An ideal gas can radiate either a continuous black-body spectrum or a line spectrum, depending upon the density of the gas cloud. A tenuous gas has negligible reactions between its constituent atoms, and so the predominant way in which radiation is either absorbed or emitted depends only upon the way in which the electrons interact with the surrounding photons. This leads to emission and absorption spectra. In a dense gas, the energy levels are altered by the proximity of the atoms to one another, and this smears the radiation out into a continuous band, producing a continuum.

Spectral lines have a characteristic shape known as a Gaussian profile (see Figure

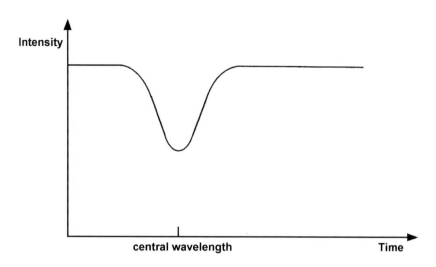

Figure 2.7. A Gaussian line profile. The precise shape is determined by the physical conditions of the gas in which the line was produced. Temperature, density and pressure are all factors which will affect the width of the line. This image is the profile of an absorption line. The higher surroundings of the Gaussian profile are called the continuum, and correspond to the continuous spectrum upon which the absorption line is superimposed.

2.7). This is because there is a natural tendency for the lines to spread out over a small range of wavelengths. For the atoms in an ideal gas, the temperature of the cloud will translate directly into a velocity distribution for its constituent atoms. When an electron absorbs and then re-emits a photon of radiation, the wavelength of the photon will be slightly altered by the individual motion of the emitting atom. When the cloud is studied as a whole, the statistical distribution of velocities (described by Maxwell) creates these Gaussian profiles. The process is known as thermal Doppler broadening. In the case of turbulent motion within the cloud being studied, turbulent Doppler broadening will also take place. The analysis of the shape of spectral lines can therefore tell us what physical conditions are possessed by the emitting (or absorbing) cloud.

As the density of the cloud increases, the atoms become more crowded and the amount of interaction between them becomes significant. Our first clue can be found by remembering that one of our conditions for black-body radiation is that the cloud must be in thermal equilibrium. This implies that there must be a large number of atomic interactions in order to spread the thermal energy evenly between atoms in the gas. In the course of these collisions, the atoms will interact with each other. This occurs because the wave nature of the electron allows it to act as if its charge is distributed evenly about the atomic nucleus. Thus the region in which the particulate electron must be located is known as the electron cloud. The atoms in a dense ideal gas which is in thermal equilibrium interact by feeling a repulsive force between their surrounding electron clouds. This alters the motion of the interacting electrons, causing a corresponding change to the energy of the system. That energy is lost in the form of a photon. For our purposes we can think of the energies involved in these processes as being unquantised, and so the radiation is emitted in a continuum which, because of the distribution of atomic energies, produces a Planck curve.

The careful analysis of the spectrum of any celestial object can therefore reveal a great deal about the matter contained within that object. The shape of the spectrum, the presence or absence of spectral lines and the intensity of radiation all help in understanding the physical processes by which the radiation was emitted.

2.3   ELECTROMAGNETIC RADIATION EMISSION MECHANISMS

Having examined thermal emission, it is now important to consider the other emission mechanisms of electromagnetic radiation. Considering bound–bound transitions, it becomes natural to wonder if it is possible for electrons to completely escape the influence of their atomic nuclei simply by absorbing photons of sufficiently high energy. This is a process known as photoionisation, and is defined as a bound–free transition.

The energy required to ionise a hydrogen atom, if its electron is in the ground state, is 13.6 eV. (An electron volt is a measure of the energy gained by an electron if it is accelerated through a potential difference of 1 volt). Photons containing this amount of energy or more can be absorbed and used to eject the electron from the atom, leaving an atomic nucleus with a net positive charge. Energy exceeding the 13.6 eV required for ionisation will be converted into kinetic energy by the free electron. As the photon energy increases past the ionisation limit, however, the probability that the photon will be absorbed decreases, and so absorption bands are created which have a sharp discontinuity corresponding to the energy of ionisation (see Figure 2.8). Several such bands are possible within the same spectrum because ionisation can take place from electrons in any energy level. A gas cloud which is largely composed of positive ions and electrons is known as a plasma.

Conversely, free electrons can be captured by positive ions and their energy given out as photons. This is the process of recombination, and is known as a free–bound transition. The continual photoionisation of atoms and their recombination is the process which gives rise to glowing clouds of gas known as emission nebulae, which surround young high-mass stars (see Figure 2.9).

The final type of emission mechanism to be considered is a range of mechanisms which could be classed as free–free transitions. In all of these interactions the electrons are unbound to atomic nuclei, and remain that way, even after the interaction which produces the radiation. The most obvious of the free–free transitions is known as thermal bremsstrahlung. This occurs when a free electron is decelerated by an interaction with another charged particle. The other particle may be a positive ion (for example, an ionised atomic nucleus) or another free electron. Whatever the precise situation, the energy lost by the electron in the interaction is released as a single photon of radiation. A thermal bremsstrahlung spectrum is

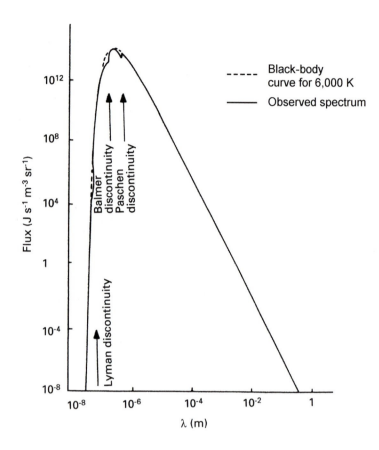

Figure 2.8. Schematic continuous spectrum of a solar-type (G-type) star showing ionisation bands. (Adapted from Kitchin, C., *Stars, Nebulae and the Interstellar Medium*, Adam Hilger, 1987.)

continuous, but of a very different shape from a black-body curve. It is the characteristic emission of a tenuous plasma.

Magnetobremsstrahlung is also possible, and is usually termed synchrotron radiation. This emission mechanism involves particles travelling at relativistic velocities in circular trajectories through a magnetic field. This causes the electron to lose energy by emitting electromagnetic radiation. Unlike other emission mechanisms, the radiation produced in this way is highly polarised. This means that instead of being randomly oriented, the electric vectors of the photons are consistently placed in the same direction.

The Compton effect occurs when a high-energy photon interacts with a lower-energy electron. It is convenient to imagine that the photon is scattered and knocked into a lower energy state, whilst the electron is boosted to higher energies. The inverse process occurs when a high-energy electron interacts with a low-energy

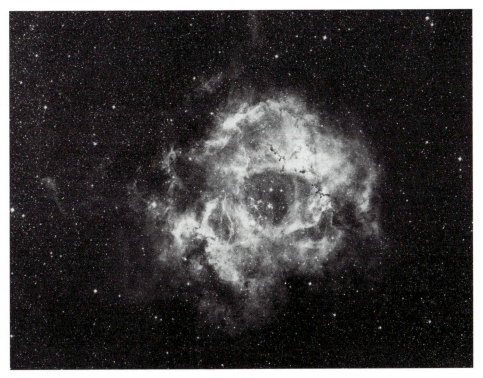

Figure 2.9. The Rosette nebula. High-mass stars ionise the surrounding hydrogen gas, causing emission nebulae to form around them. (Photograph reproduced courtesy of AURA/NOAO/NSF.)

photon and boosts its energy. Described in this way it appears to be rather similar to synchrotron radiation, with a photon field replacing the magnetic field. Although we have referred to these effects taking place between electrons and photons – which will apply in the majority of astrophysical cases – the electrons could be replaced by other particles.

The wavelength shift, $\Delta\lambda$, imparted on a photon in the inverse Compton effect is defined by the equation:

$$\Delta\lambda = \lambda_c\,(1- \cos\varphi) \tag{2.6}$$

where φ is the scattering angle of the photon and λ_c is the Compton wavelength of the particle subjected to the photon interaction. The Compton wavelength is the wavelength a photon would possess if it carried the rest energy of the particle; hence it is defined:

$$\lambda_c = \frac{h}{m_0 c} \tag{2.7}$$

where m_0 is the mass of the particle involved in the scattering, h is the Planck constant, and c is the speed of light.

2.4   WINDOWS ON THE UNIVERSE

As the previous section has shown, different types of emission mechanism are produced in response to different physical conditions. Within the boundaries of each emission mechanism, a wide range of radiation wavelengths can also be produced. Observing the Universe at different wavelengths and scrutinising spectra can therefore provide an insight into the different phenomena and physical environments present throughout the cosmos. In some ways, we have become so accustomed to seeing the night sky at optical wavelengths, that spacecraft images of it in anything other than visible light often take us by surprise. However, it is important to remember at all times that visible light is such a tiny part of the electromagnetic spectrum that to ignore the rest would be folly.

If we begin at the lowest energies of electromagnetic radiation, the Universe becomes the realm of the radio galaxies. The entire sky is dotted by their tremendously powerful radio-emitting lobes. Also 'visible' at radio wavelengths are objects from our own Galaxy known as supernova remnants. These are the glowing remains of exploded high-mass stars. In the direction of Sagittarius, the centre of the Milky Way would glow brilliantly. At certain radio wavelengths, vast clouds of molecular gas can be seen. These also exist within our own Galaxy, and are the clouds out of which stars eventually form. The different wavelengths of radio trace out different molecular species. By mapping the strength of each molecule's radio emission, contour maps of the clouds can be produced, which show the number of each type of molecule at each point throughout the cloud. The nearest molecular clouds would, from our viewpoint, appear enormous. For instance, the molecular cloud with which the Orion nebula is associated encompasses the whole of the constellation at radio wavelengths.

At the next man-made division – microwave wavelengths – the view is totally different, since the cosmic microwave background dominates. Instead of a dark sky punctuated with both point and extended sources, the entire sky is bright and glowing with energy. Superficially, the brightness appears the same in all directions. There are variations in the radiation, however, and the first, most obvious, is known as the dipolar anisotropy. It is produced by the motion of the Earth relative to the cosmic microwave background radiation (CMBR). When all the components of the Earth's motions – such as those due to the Solar System's motion around the centre of our Galaxy, and the Galaxy's motion within the Local Group – are summed, the resultant velocity increases the temperature of the background CMBR in the direction of the Earth's motion (see Figure 2.10). It also decreases its temperature by a corresponding amount in the antithetical direction. Underlying this is the microwave contribution from the material and objects in the Milky Way. If all of these are removed and the remaining microwaves subjected to very careful analysis, fluctuations which coincide with the emergence of Galactic structure would become visible.

Changing our observations to the more energetic bands of radiation, we arrive at the infrared. The sky is once again dark, and stars have appeared on the scene. In general these are the stars which are cooler than can be easily seen at optical wavelengths. They are the red dwarfs and red giants. Also present are the nascent

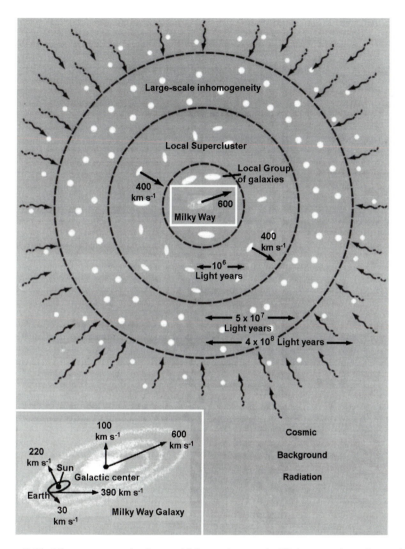

Figure 2.10. The vantage point from which we observe the Universe at large is anything but static. Our frame of reference is a superposition of many different motions. Our overall velocity can be measured with reference to the cosmic microwave background radiation, because of the Doppler shift that the velocity induces. (Adapted from Silk, J., *A Short History of the Universe*, W.H. Freeman, 1994.)

stars in stellar nurseries such as the Orion nebula. These protostars and other young stellar objects are obscured from view at optical wavelengths by their dusty birth clouds, but at infrared wavelengths they shine through. The Milky Way continues to stretch across the sky again, illuminated not so much by stars but by the infrared glow of warm dust clouds. A second band also stretches across the sky, intersecting the Milky Way at an angle of slightly less than 80 degrees and marking the plane of

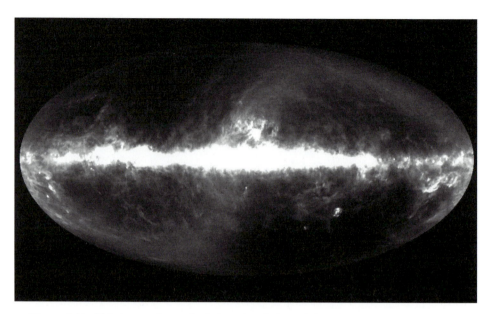

Figure 2.11. This image shows the infrared emission of dust from the Galaxy running horizontally across the centre. The wisps of emission above and below this are termed the infrared cirrus, and also emanate from dust in our Galaxy. The sweeping S-shaped band that appears superimposed upon the Galactic emission is dust lying within our Solar System – the zodiacal dust. It is confined to the plane of the ecliptic but, because of the projection used to map the whole sky onto this ellipse, it has been distorted into the S-shape. (Photograph reproduced courtesy Michael Hauser (Space Telescope Science Institute), the COBE/DIRBE Science Team and NASA.)

our Solar System. It is glowing because there is warm dust in interplanetary space (see Figure 2.11).

Beyond infrared, the familiar optical spectrum occurs. (Our view of the Universe at these wavelengths is described in Chapter 1.) At higher energies – the ultraviolet – the very hottest stars in the Galaxy dominate the view. They are the O- and B-type supergiant stars which have surface temperatures of 40,000 K or more. Using Wien's law, λ_{max} can be calculated to be 70 nm – a wavelength in the extreme ultraviolet region of the spectrum. Thus, whereas most stars would fade to obscurity, the brightest stars in the Galaxy would become even brighter if we could see the ultraviolet region of the spectrum.

As well as the individual stars, there is also a glow from all around. This is known as the local bubble, and is a roughly spherical 'cavity' within which the Solar System and surrounding stars sit. It is thought to have been produced by a supernova explosion long ago. The shock wave from that explosion has swept interstellar atoms into a bubble shape which now glows softly at ultraviolet wavelengths.

Passing on to even more energetic wavelengths, the sky continues to glow faintly at X-ray energies. Unlike the locally produced ultraviolet, the X-ray background emanates from vast distances, and is thought to be the combined emission from distant galaxies. Clusters of galaxies glow faintly in X-rays due to thermal bremsstrahlung, as do other regions of hot, tenuous plasmas present in our own Galaxy. This general background glow is also punctuated by point sources of intense X-rays. Some of these are the nuclei of powerful active galaxies. Others mark the positions in our Galaxy where stars are being ripped to pieces by incredibly strong gravitational fields. Both phenomena are explained theoretically as being due to the action of black holes.

If we were to continue to the highest energy range of the electromagnetic spectrum, the night sky would look perfectly black again except for the occasional flash of an errant gamma-ray. Until recently, these bursts – which occur once every day or so from totally random directions – were unexplained. Now progress has been made, and it is currently thought that they are caused by the explosion of very massive stars in galaxies that are at vast distances.

Our view of the cosmos would therefore be very biased if we were to persist in myopically observing it at optical wavelengths alone. In a perfect world, astronomers would be able to tune a telescope to detect radiation at wavelengths of their own choosing, but there are certain mitigating factors. Different focusing devices and detectors are needed, for example, in order to receive different types of radiation. And another (far more serious) problem is that even if a detector for each type of radiation were to be built and mounted, only some of them would receive signals. This is because the Earth's atmosphere prevents certain types of radiation from reaching the surface of our world (see Figure 2.12).

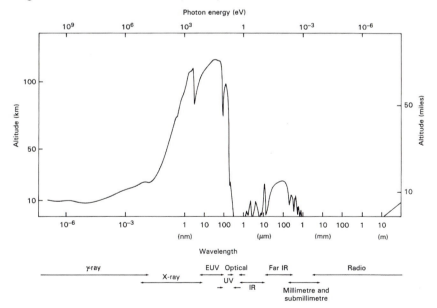

Figure 2.12. Not all electromagnetic radiation is capable of penetrating the Earth's atmosphere. The ozone layer absorbs a large amount of high-energy radiation and molecules in the atmosphere – particularly water vapour – block much of the infrared.

At wavelengths longer than 20 m, radio waves are reflected by the Earth's ionosphere. Most, but not all, of the infrared radiation is absorbed by molecules in the atmosphere. Optical wavelengths, obviously, pass through without hindrance. Apart from the ultraviolet radiation which causes suntan, all the high-energy photons are blocked by various interactions with atoms in the atmosphere. This occurs at altitudes between a few tens to a few hundreds of kilometres. The type of radiation we wish to observe determines the style of detector which has to be built and whether or not it has to be made into a satellite and sent into space.

2.5   TELESCOPES, DETECTORS AND SPACE PROBES

From Earth-based observatories astronomers can detect predominantly optical and radio waves. The use of telescopes for optical astronomy began with Galileo in 1609. He applied the design of Hans Lippershey, a Dutch spectacle maker, and improved it. His telescope allowed him to make the most remarkable discoveries of his age, including observations of the moons of Jupiter which strengthened the Copernican view of a heliocentric Solar System. It also encouraged those who followed him to build bigger and better telescopes in an effort to discover more and more. Newton himself made a type of reflecting telescope which is still widely used today by amateur astronomers, and is known appropriately as the Newtonian.

Today's professional astronomical telescopes are enormous compared with those original ones. Many existing telescopes have mirrors which are 4 m in diameter, whilst a new generation of telescopes are just coming on line which possess 8-m mirrors. The largest telescopes in the world are the twin Keck telescopes on the extinct volcano, Mauna Kea, Hawaii, each of which possesses a mirror measuring 10 m across. These mirrors are so large that it was unrealistic to build two single, continuous mirrors of that size. Instead, each one is constructed like an insect's compound eye, with 36 hexagonal mirror segments, held in precise place by electronic support arms. Following the building of the first Keck telescope, it proved to be so successful that a second, identical telescope was constructed alongside it. (see Figure 2.13). This siting of the telescopes is not unusual for modern observatories, despite the sometimes difficult conditions of working at altitude. Mountain tops are ideal locations, because the telescopes are elevated above most of the weather and turbulent air which obstructs observations from sea level.

Atmospheric turbulence can also be combated in other ways. For example, a technology known as active and adaptive optics is just proving itself viable. Active and adaptive optics form a system which allows a reflecting surface to be controlled by mechanical actuators. The quality of the image being produced by the telescope is continuously monitored, any deviation from perfection is registered, and corrections are incorporated by a computer, which determines how the mirror should be adjusted to produce a perfect image. (see Figure 2.14). This system has two advantages, the first of which is that a large mirror can be manufactured and supported in many places. Formerly, the rigid construction of a mirror to a level which prevented flexure made it prohibitively heavy. The second advantage is that,

Figure 2.13. Some of the largest telescopes in the world are situated on Mauna Kea. At an altitude of 4 km, they are well above most of the clouds that plague observatories at or near sea-level. (Photograph reproduced courtesy of Richard Wainscoat/Gemini Observatory/AURA/NSF.)

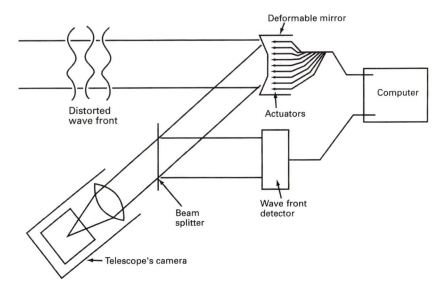

Figure 2.14. Schematic of an adaptive optics system. The key component is the wave front analyser, which senses the distortion in the incoming wave front, allowing the computer to adjust the mirror to compensate.

providing the system can react fast enough it can detect imperfections in the image caused by atmospheric turbulence and compensate for them by manipulating the mirror. This type of technology is currently being installed in more and more observatories.

A telescope's primary function is the collection of light, and it is the task of the astronomer to decide what is done with that light once it has been collected. Originally, visual observations were sufficient, but contemporary science demands hard facts and data to back up assertions, and a plethora of techniques and detectors is now available to enable astronomers to record their findings.

Two basic techniques of analysing the light collected by telescopes are those of spectroscopy and photometry. A spectroscope splits the incoming light into a spectrum so that its spectral lines can be studied. As discussed earlier, the pattern of spectral lines can indicate the chemicals present in the object under investigation, and the size and shape of the spectral lines can lead to conclusions about the motion of that object. Above and below the spectrum, the spectroscope superimposes reference emission lines, so that accurate wavelengths for the observed spectral lines can be measured. Formerly, microdensitometers were used to study the intensity of radiation along the spectrum on a photographic film, and it was from those traces that the shapes of the spectral lines could be determined. Today, computers are used to manipulate the spectral data and produce such traces (see Figure 2.15).

It is fair to say that the vast majority of astronomers use spectroscopy to study their chosen objects because the level of information returned from a single observation can be very high. However, the recording of spectra can be time-consuming, because with traditional spectroscopes the light from an object is passed

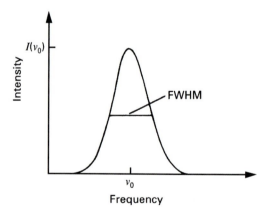

Figure 2.15. Typical trace across a spectral emission line. The Gaussian shape of spectral lines can be obtained in a number of ways. When imaging spectra with photographic plates, they can be obtained by tracing across a spectral line with a microdensitometer. Nowadays, spectra are recorded by CCDs, and the data are digitised and stored in a computer, which allows software to quickly and easily display the Gaussian profiles. (Adapted from Kitchin, C., *Stars, Nebulae and the Interstellar Medium*, Adam Hilger, 1987.)

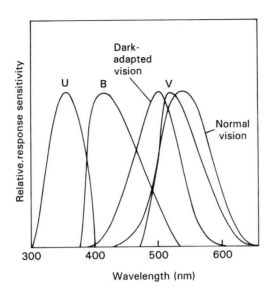

Figure 2.16. The spectral response of the dark-adapted eye and of the U (ultraviolet) centred on 265 nm, B (blue) centred on 440 nm, and V (visual – yellow–green) centred on 550 nm, pass-band filters used for photometry in the UBV system.

through a slit and then expanded into its constituent wavelengths. This can dramatically decrease its brightness and thus increase the imaging time necessary for the exposure.

Photometry is a technique whereby the brightness of a celestial object is recorded at different wavelengths. These wavelengths are known as photometric bands, and have been standardised to specific wavelengths. They have also been assigned letters: for example, the K photometric band refers to the wavelength of 2.22 μm, which is in the infrared region of the spectrum. The technique is practised by placing filters between the telescope and some form of photon-counting device. The filters allow only specific wavelengths of light through to the counter, which records the intensity of light at that wavelength (see Figure 2.16).

Photometry can be used to classify stars because it can reveal their black-body temperatures. It is very useful at tracking variations in the amount of radiation which is output by celestial objects over a period of time. A good example of where photometry is particularly useful is in the classification of variable stars. Photometry reveals just how their light output varies over the course of time, and allows light-curves to be constructed (see Figure 2.17). Sophisticated analysis of photometric data covering a wide range of wavelengths can yield many of the same results derived from spectroscopy; but, unfortunately, the collection of photometric data is even more time-consuming than the collection of spectroscopic data. Thus, the construction of highly complicated spectroscopes is still favoured over the use of the simpler photometers.

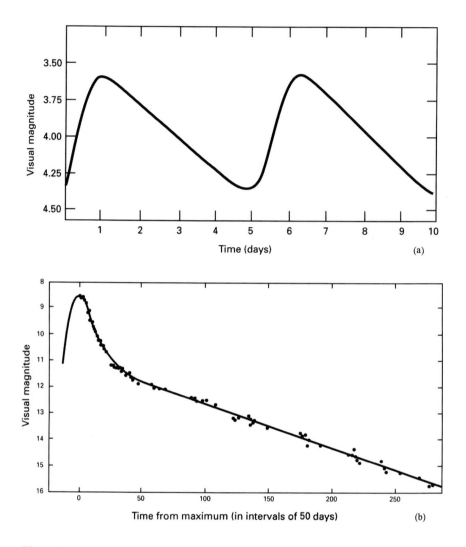

Figure 2.17. The light-curve of (a) a Cepheid variable or (b) a supernova explosion can provide excellent diagnostic information for the astronomers studying these objects.

A third – less popular, though no less valid – technique is that of polarimetry. The basis of this method of observing celestial objects is the collection of information about the orientation of the electric vector in the electromagnetic radiation being collected by the telescope. It can be a powerful tool in probing regions where radiation is being scattered by dust or electrons, and can also help to reveal regions of space containing magnetic fields which have aligned dust grains. Thus, the technique is excellent for probing the interstellar medium in our Galaxy and in others.

Visual observations are now insufficient for the professional astronomer. The light captured by large telescopes and processed via one of the above techniques must be recorded for later analysis and publication. Photography is still used in certain isolated cases, but by far the most rapidly developing technology is the CCD (charge-coupled device), which is so sensitive that it records almost every photon which strikes it. A CCD is a computer chip which uses a detection mechanism similar to the photoelectric effect to record the number of photons striking it. The photons are instantly converted into an electrical signal, which can be read out into a computer for later analysis and display.

A CCD's ability to count almost every photon which strikes it is referred to as its high quantum efficiency. This is a ratio which can be easily understood as the number of detected photons divided by the number of incident photons; in other words, the percentage of photons which are actually detected when they strike the detector. A CCD will typically have a quantum efficiency in excess of 75%, whereas a photographic system will rarely exceed 1%.

CCD chips are made of semiconductors, the choice of which determines which part of the electromagnetic spectrum it will detect. Ultraviolet, optical and infrared can now all be observed with the correct CCD. The collection of radio waves, however, is undertaken with totally different types of telescope and detector.

A radio dish works in exactly the same way as an optical telescope. The reflector still needs to be parabolic, but because the wavelengths of radio radiation are so much longer it does not need to be aluminised. The detector is placed in the prime focus position on the radio dish, where a secondary mirror is usually (but not always) positioned on an optical telescope.

Radio astronomy has pioneered a technique known as interferometry, which effectively increases the resolving power of a telescope by linking it in tandem with another. The resolving power of a telescope is a measure of how apparently close two celestial objects can be before the telescope fails to show them as two distinct objects. For optical telescopes, the resolving power is often so good that the distortion of the images through atmospheric turbulence is the biggest source of image degradation. For radio telescopes this is not the case, because radio wavelengths are so much longer.

The resolving power of a telescope is defined as the wavelength of radiation divided by the aperture of the telescope. Thus, with an increase in the wavelength, the resolving power of the telescope can be maintained only with a commensurate increase in the aperture of the telescope. The radio waves emitted by neutral hydrogen are six orders of magnitude longer than visible light; so, in order to achieve the same resolution from a radio telescope the aperture would need to be one million times greater than that of an optical telescope! Obviously, another method had to be devised!

The problem was solved with the introduction of interferometry, a technique pioneered at Cambridge University in the 1940s. It allows an object to be observed simultaneously by two or more radio telescopes. The signals collected by the dishes are then combined and interference patterns are obtained. Using Fourier analysis, the interference patterns can be used to construct detailed images with much greater resolution than can be obtained with a single dish. Interferometry in the radio region of the spectrum has been so successful that the method has now also been applied to optical telescopes. A pioneering team of astronomers at Cambridge has applied the technology to produce an image of the star Capella using four small telescopes together. They have called their venture COAST – Cambridge Optical Aperture Synthesis Telescope. There are now under construction, around the world, a number of much larger observatories which all hope to exploit optical interferometry. Two prime examples are the four 8-m telescopes of the European Southern Observatory's Very Large Telescope (VLT) and the two 10-m Keck telescopes.

The impenetrability of the Earth's atmosphere makes it necessary to send space probes and satellites into orbit in order to collect those wavelengths from which we are shielded. High-energy radiation is almost totally blocked out by the atmosphere. Apart from the small amount of ultraviolet radiation which penetrates the atmosphere and causes suntan, the vast majority of ultraviolet radiation has to be collected by space probes such as the International Ultraviolet Explorer (IUE). At even shorter wavelengths, the Einstein X-Ray Observatory has surveyed that region of the electromagnetic spectrum. These space probes require a slightly different type of mirror system, known as a grazing incidence mirror. This type of mirror is an annulus taken from the cylindrical wall of a paraboloid rather than taken from the bowl as for optical telescopes. In order to maximise the amount of high-energy photons collected, annuli of successively smaller radii are nested within one another.

Molecular absorption blocks a large amount of infrared radiation, but this too has now been sampled from space by the Infrared Astronomical Satellite (IRAS) and the Infrared Space Observatory (ISO). Even wavelengths which can be studied from Earth have been collected in space. High above the Earth's atmosphere, the star images remain steady and unwaveringly sharp. The celebrated Hubble Space Telescope (HST), although occasionally requiring upgrades and servicing, is currently performing magnificently by collecting unprecedented images of the cosmos.

2.6   OTHER METHODS OF OBSERVING THE UNIVERSE

Whilst it is true to say that 90% of observations are undertaken by detecting electromagnetic radiation from space, there are other emissions which can be studied. One method is the detection of tiny particles known as neutrinos. These particles are a fundamental constituent of the cosmos and carry excess energy away from nuclear reactions. Neutrino detectors have been very successful in detecting these fleeting particles after they have been produced in the heart of the Sun. Additionally – and perhaps their greatest collective claim to fame – the detectors identified a burst of neutrinos from the 'nearby' supernova of 1987, exactly as predicted by theory.

Undeniably, neutrino detectors are still very crude when compared with electromagnetic detectors. They cannot, for example, determine from which area of the sky the neutrinos emanate. However, the continual refinement of these devices will provide another powerful resource with which to probe the Universe in ever-increasing detail.

Another technology of instrumentation which is in its infancy at the moment is the gravitational radiation detector. Gravitational radiation is predicted, by the general theory of relativity, to be released from massive bodies which are changing their gravitational relationship with one another. It would propagate through the spacetime continuum in the form of a longitudinal wave motion (a repeating pattern of compressions and rarefactions) and distort any objects in its path. The distortions would manifest themselves as minuscule variations in the length of objects. As yet, however, technology is insufficiently advanced to be able to detect these waves, since the size of the variations they induce are exceedingly small percentages of the diameter of a hydrogen atom. Meanwhile, research is forever advancing and the sensitivity is gradually being pushed closer and closer to the detection limit.

Gravitational radiation is the final test of the theory of general relativity. There is even a hope that it may be possible to detect a cosmic gravitational background radiation similar to the cosmic microwave background radiation. If this were to prove possible, then theoretically it would afford cosmologists a view of the Universe as it existed one millionth of a millionth of a second after the Big Bang!

3

The architecture of matter

The concept that the Universe is made up of small constituents, called atoms, dates back over 2,600 years, to the Greeks. Half a millennium after the idea first formed in the Grecian mind, the poet Lucretius wrote a book called *On the Nature of Things*, in which he presented the 'atomic' ideas of Leucippus and Democritus. It was a sophisticated argument for the existence of atoms. Certainly, the details concerning the real nature of an atom were missing (they would be established by twentieth-century science), but the basic concept was firmly in place. It is interesting to note that in Greek scientific circles a rival to the atom hypothesis was that the constituents of nature were strings – singular particles that produced strings when one thought of them moving through time. This thinking foreshadowed the work of Einstein in his linking of space and time in his general theory of relativity, but in ancient Greece the idea was abandoned in favour of atoms.

The manuscript of Lucretius was discovered as an archaeological find in 1417. Copies were subsequently produced following the introduction of the printing press in the 1450s, and in the following centuries scientists and natural philosophers felt compelled to follow up the work of their ancient Greek predecessors. This work has led to our modern understanding of particles – a contemporary edifice of abstract reason that is itself as remarkable as the Big Bang theory.

3.1 THE RUTHERFORD ATOM

From the fifteenth century onwards, more and more scientists became intrigued by the notion of atoms; and more and more of them set out to prove that atoms existed. Hitherto unexplainable phenomena began to make sense if matter – particularly air – was thought to be made of atoms. In 1808, English chemist John Dalton amalgamated the known experimental data in a remarkable book in which he showed not only that was matter made of atoms, but that there could be different types of atom, with different chemical properties. With this work, the concept of the chemical element was firmly introduced into science. In the 1870s, Russian scientist Dimitri Mendeleev succeeded in assembling the chemical elements in order by creating the periodic table. This tool allowed him to predict the existence of new chemical elements. Once they had been discovered, most scientists accepted the existence of atoms and elements.

The flurry of scientific interest in chemical elements led to the discovery of radioactivity in 1895 – by the French physicist Antoine Becquerel – and the suggestion that atoms were themselves made of smaller particles. It was also deduced that the particles carried electrical charges, either positive or negative. In the 1890s Joseph Thomson, of Cambridge University, discovered the negatively charged particle – the electron. He proposed that the structure of the atom was such that the positive charge was spread uniformly throughout its volume, with the negatively charged electrons distributed discretely inside the atom in the same way as the fruit in a plum pudding.

However, in 1911 Ernest Rutherford demonstrated the results of an experiment which could not be understood by the 'plum-pudding' approach. He suggested to colleagues H. Geiger and E. Marsden that they direct a beam of radioactive particles onto a thin sheet of gold and count how many particles pass straight through the sheet. This would enable them to determine how tightly packed were the gold atoms in the sheet. Their results were amazing: almost all of the particles passed straight through as if the gold sheet did not exist. Some were deflected slightly, indicating that they had suffered close interactions with the gold atoms, but only a small proportion actually rebounded, as would be expected if they collided with the gold atoms (see Figure 3.1).

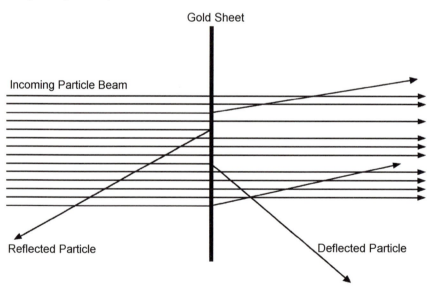

Figure 3.1. Rutherford's discovery of the atomic nucleus. By firing radioactive particles (α particles) at a thin sheet of gold, Rutherford and his colleagues discovered that most pass straight through the sheet. Only a few are deflected, and even fewer rebound. This is feasible only if most of the gold sheet is empty space. The mass of each individual gold atom is concentrated into a tiny volume at the centre of each atom – the nucleus.

Rutherford's conclusion was inescapable: most of an atom is empty space. The vast majority of its mass and all of its positive electrical charge is tightly concentrated in a nucleus at the centre of the atom's sphere of influence, and the negatively charged particles orbit this nucleus. The Rutherford experiment proved that the atomic nucleus is 10,000 times smaller than the atom itself. Later, Rutherford showed that atomic nuclei contain positively charged particles that are identical to the nuclei of hydrogen atoms. These he named them positrons. According to the theory, a proton carries a positive charge, and an electron carries a negative charge.

Since the time of Rutherford, experiments and theory have driven the understanding of the nature of the atom. The advent of quantum theory has driven research to deeper levels of knowledge about atomic nuclei. In one early example, quantum theory allowed the explanation of the chemical properties of matter to reveal an obvious discrepancy between the number of protons contained in and the mass of most nuclei. In 1926, it led to the realisation that there must exist another particle, similar to a proton and yet carrying no electrical charge. The concept of the neutron was born.

Quantum theory also showed that protons and neutrons could not be singular particles, but must in turn be composed of something. The analysis proceeded in a similar way to that used by Rutherford to show that an atom was not a continuous singular entity, but something that contained a central, tightly packed point.

The magnitude of the electrical field at a surface can be calculated using Gauss' law. If the surface is a sphere then Gauss' law becomes

$$E = \frac{1}{4\pi\varepsilon_0} \frac{q}{r^2} \tag{3.1}$$

where q is the charge that the sphere surrounds, r is the radius of the sphere, and ε_0 is the permittivity of free space and quantifies the ease with which the effects of an electric charge can be communicated through a vacuum. It is a universal constant, and possesses the value 8.854×10^{-12} F/m.

The crucial aspect of equation (3.1) is the r^2 term. The larger the sphere's radius, the smaller the magnitude of the electric field. If the Gaussian sphere is defined to be the extent of the positive charge, then in the case of the 'plum-pudding' model of the atoms with $r \sim 10^{-10}$ m, the magnitude of the electric field will be much smaller than in the case of Rutherford's nucleus model with $r \sim 10^{-15}$ m.

Similarly, the proton models predicted too weak an electric field, and caused physicists to wonder if the proton itself were a composite particle made up of smaller positive charge-carrying particles. But how could such an assertion be proved?

3.2   PARTICLE ACCELERATORS

The only way to really determine whether a proton is a composite particle is to fire smaller particles into it and measure how many emerge undeflected on the other side. This is exactly the same principle that Rutherford used when investigating the gold atoms. In the case of protons, the much smaller and lighter particles, electrons, fitted the bill. If the electrons could be sufficiently accelerated they would be able to penetrate the protons.

The electrons, being charged particles, can be accelerated by placing them in an electric field, and their direction of motion can be altered by the use of magnetic fields. In the early days of particle accelerators – just after the end of World War II – simple linear accelerators were produced. The Stanford Linear Accelerator Center (SLAC) was built to probe the architecture of protons by colliding electrons with them. The results were promising, and appeared to show that protons were indeed made up of much smaller particles. In these experiments, and in others, it also became apparent that other particles, which shared some common characteristics of protons and neutrons, could be produced. These similar particles were grouped together and called hadrons.

As more hadrons were discovered, many physicists realised that a small number of fundamental building blocks could be arranged in a variety of ways in order to produce the hadrons being seen. Their building blocks were called quarks. In 1968, SLAC was sufficiently upgraded to prove beyond doubt that protons – and by extension, all hadrons – were composed of quarks. In short order the theorists had begun to understand the behaviour of particles by reducing everything down to its bare mathematical bones. This set of mathematical laws is known as the Standard Model of particle physics. With it, exponents of the theory could predict the results of particle interactions. All that was needed to test the theory's predictions were more sophisticated particle accelerators.

By 'more sophisticated', we really mean 'more powerful'. The more energy that can be carried into a reaction, the greater the number and/or the more massive the particles that can be produced. This is a direct application of

$$E = mc^2 \qquad\qquad (3.2)$$

which we first met as equation (1.2) in the concept of nuclear fusion turning mass into energy. In particle accelerators, the mass of the colliding particles is transformed into energy which combines with the kinetic energy carried by the particles, to be spontaneously converted into new particles. The probabilities of which particles should be produced from the collision of two particles at high energy can be derived from the Standard Model.

It is here that the particle physics begins to mesh almost indistinguishably with cosmology. Collision energies can now be generated that are so large that the conditions they create are unlike anything which occurs on a large scale in the Universe today. Billions of years ago – during the first three minutes of the Big Bang – conditions were very different, and were almost certainly similar to those generated inside particle accelerators.

The study of particle physics is the perfect synthesis of the very small – sub-atomic particles – with the very big – the entire Universe. Inside particle accelerators, streams of sub-atomic particles are accelerated to velocities close to the speed of light, which supplies them with similar quantities of energy that they would have possessed during the early period of the Universe. To simulate the density of matter of those early times, highly accelerated beams of particles are then collided with each other. The annihilations which result cause a liberation of energy which, in turn, leads to the creation of particles which must have populated the early Universe in large numbers but no longer exist naturally in our present-day low-energy cosmos. So, by probing the architecture of matter, physicists cannot avoid becoming cosmologists.

Today's particle accelerators are colossal monsters. As they become ever more powerful, they are able to probe ever more energetic interactions. Historically, increases in experimental energy have brought with them the discovery of new particles.

The detectors used in particle accelerators are generally all built to a standard, optimised design. Two streams of particles moving in opposite directions are directed along the axis of the cylindrical detector, where they collide head-on with each other. The energy liberated in the collision almost immediately materialises as a cascade of particles that race away from the collision point through the detector. The detector itself consists of four layers, each designed to detect a different type of particle. The layer closest to the collision point is called the tracking chamber. Every particle travelling through the tracking chamber that possesses an electrical charge will be detected, and its passage through the chamber tracked. The next layer is known as the electromagnetic calorimeter. This is specially designed to not only track particles with electric charge but also to induce them to lose energy. The lighter the particle, the more easily it will lose energy, and so particles such as electrons will actually be halted in the electromagnetic calorimeter, as will photons.

The heavier charged particles, such as protons, will not stop, and they will pass into the next layer of the detector – the hadronic calorimeter. They will finally be forced to a halt, so their tracks will stop in this layer of the detector. The only type of particle that will still be moving by this stage is the muon, which can be considered to be a heavy 'cousin' of the electron. The fourth and final detector layer is therefore called the muon chamber (see Figure 3.2).

In addition to these layers, a strong magnetic field is generated around the detector. This causes the electrically charged particles' tracks to curve. The precise amount of the curvature can be used to deduce the electrical charge and mass of the particle, thus aiding its identification.

As well as the Stanford accelerator (mentioned at the beginning of this chapter) there are, around the world, other accelerators performing crucial experiments to test both the Standard Model and the Big Bang. These include, in the USA, the Fermi National Accelerator Laboratory (Fermilab) near Chicago, and the Brookhaven National Laboratory on Long Island. Europe's premier particle accelerator is CERN, which straddles the Franco-Swiss border (see Figure 3.3).

3.3   THE STANDARD MODEL

The Standard Model is used today to describe almost all particles and their interactions. According to the theory, the fundamental particles of nature which

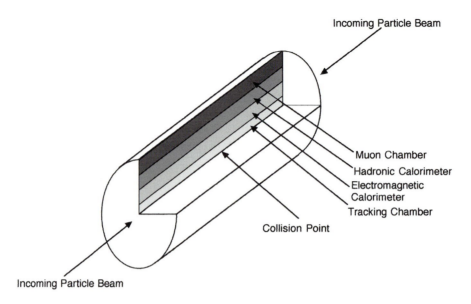

Figure 3.2. Schematic of a particle accelerator detector. Streams of oppositely directed particles are fed into the detector so that they collide in the middle. The particles created in the collision then cascade outwards into the detector, where they can be measured. Some decay into other particles so quickly that they do not even enter the detector's chambers, in which case the decay products are detected and used to infer which particles must have been produced as a direct result of the collisions.

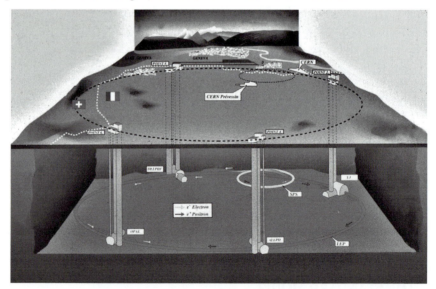

Figure 3.3. The particle accelerator at CERN. The Alps are in the background, the Geneva plain is in the centre and the underground experimental set-up is in the foreground. (Image reproduced courtesy of CERN Geneva.)

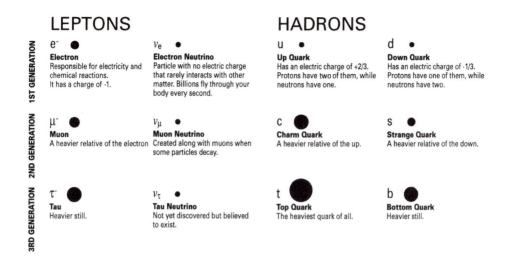

Figure 3.4. The fundamental particles of nature – the leptons and quarks – are described by particle physics' Standard Model. Both families are split into three generations, each of which contains a pair of particles. The successive generations are similar in electrical charge but different in mass. Ordinary matter mostly uses the lightest first-generation particles. (Adapted from *Big Bang Science*, PPARC.)

constitute matter fall naturally into two families, known as leptons and quarks. Both families contain six particles which are split into three generations of pairs. The most familiar leptons are the first generation pair – the electron and the electron neutrino (usually referred to simply as the neutrino). The other two generations are populated by heavier versions of both the electron (the muon and the tau) and the electron neutrino (the muon neutrino and the tau neutrino). The quark generations are composed of successively heavier versions of the up and down quarks: the charm, the strange, the truth and the beauty (also known as top and bottom) (see Figure 3.4).

These fundamental particles can combine in many ways to become the matter content of the Universe. For example, an electron has an electromagnetic charge of –1. The up quark has a charge of +2/3, whilst the down quark has a charge of –1/3. In order for the quarks to combine into the more familiar sub-atomic particles – neutrons and protons – they must come together in the correct combinations. Two up quarks and one down quark make up a proton which possesses a charge of +1. One up quark and two down quarks come together to form a neutron which carries no charge. These hybrid particles made from quarks are known as hadrons. Hadrons can be subdivided into quark triplets known as baryons, such as protons and neutrons and mesons, which consist of a quark and its antimatter counterpart (see Figure 3.5). Electrons, protons and neutrons then combine to form atoms. The exact number of protons defines the atom's chemical identity; the exact number of neutrons defines the atom's isotopic identity; and the exact number of electrons defines the atom's ionisation state.

There are very few stable particles in the Universe today. Only the electron and

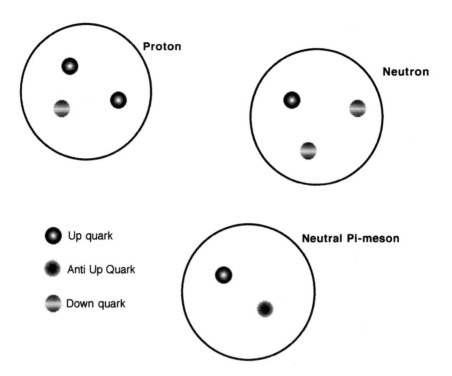

Figure 3.5. A hadron is a particle made of quarks. A quark carries a charge similar to electrical charge, except that instead of being either positive of negative, it can be any one of three states. This has been termed the 'colour' charge (although it really has nothing to do with light); red, green and blue light combine to produce white light – and every hadron has to be 'white'. Quarks can therefore be said to colour charged either red, green or blue. A baryon is a type of hadron that combines a red, green and a blue quark. Protons and neutrons are baryons. The different electrical charges carried by the quarks dictate which type of baryon is produced. There are also antimatter quarks, and so another way to create a white particle is to combine a red quark with an anti-red quark (or blue with anti-blue, and so on). These particles are called mesons, and exist only for a short time before the quark and anti-quark pair annihilates itself, producing photons.

the electron neutrino are truly stable (although photons can also be considered stable). A neutron will decay into an electron and a proton in about 896 seconds, if removed from an atomic nucleus. A proton may or may not be stable; if it is not stable it will decay in a lifetime of around 10^{31} years and become a positron and a pi-meson, which itself decays into two photons. Although their vast lifespans render protons stable for all current problems, the very long-term future of the Universe could be affected if they do decay. All the other hadrons have lifetimes of a mere fraction of a second.

All of the particles – whether they be leptons or quarks and their composites –

interact with one another via the exchange of a third group of particles known as gauge bosons. These are the particles that carry the forces of nature.

3.4   THE FUNDAMENTAL FORCES OF NATURE

Today we recognise only four forces, which we term the fundamental forces. They are gravity, electromagnetism, the weak nuclear force and the strong nuclear force, and each one of them acts differently from the others. They are carried by particles called gauge bosons in the Standard Model. The gauge boson of the electromagnetic force is the photon – a particle which we have already met. The weak nuclear force is carried by three different particles: the neutral Z^0 and the charged W^+ and W^-. These particles can be thought of as being somewhat analogous to the photon, except that the W varieties possess a charge and they all possess mass. The strong nuclear force is mediated by a gauge boson known as a gluon. This transmits the force between quarks, and virtual mesons then transmit the force between nucleons. As yet, there is no convincing quantum theory of gravity, although the hypothetical gauge boson has been named the graviton.

The mass of the gauge bosons affects the distance over which these fundamental forces can act. This is a consequence of the Heisenberg uncertainty principle, which states that

$$\Delta E \Delta t \cong \frac{h}{2\pi} \tag{3.3}$$

Since the energy required to create the gauge bosons is borrowed from the energy field of the space surrounding the interacting particles, that energy has to be paid back before equation (3.3) is violated. Equation (3.2) defines how much energy needs to be borrowed, and so the time the particle can 'live' for, Δt, is defined by a combination of the two equations:

$$\Delta t \cong \frac{h}{2\pi mc^2} \tag{3.4}$$

The range of the force, r, can therefore be shown to be inversely proportional to the mass of its gauge boson because, according to special relativity, nothing can travel faster than the speed of light in a vacuum.

$$r \leqslant c\Delta t \cong \frac{h}{2\pi mc} \; \alpha \; \frac{1}{m} \tag{3.5}$$

The weak nuclear force is carried by massive particles which limit its range to approximately 10^{-17} m. The mesons which carry the strong nuclear force are less massive, and can react over a distance of 10^{-15} m. Photons, being massless, can exist forever, and so the range of the electromagnetic force is theoretically infinite. Over large regions of space, however, electrical charge is neutral, and so the electromagnetic force is confined. Gravitons are theorised to be massless and, unlike electromagnetism – which can be positive or negative – gravity is only ever attractive, and thus cannot be cancelled out. This is the reason why it shapes the Universe on its largest scale.

Although the Standard Model is very robust, with much experimental evidence to support it, there are some inconsistencies. It is a widely held belief that these provide the route forward to extend the Standard Model into a more complete theory. For example: one of the problems that the Standard Model evades is why some particles possess mass whilst others do not.

In an extension of the Standard Model, Peter Higgs, of the University of Edinburgh, has formulated a possible solution of this problem. He has postulated a fifth force-field, which has been named after him. The Higgs field permeates the Universe, and is not like the other forces of nature. For example: an electromagnetic field is produced by the presence of a particle carrying an electric charge; remove the particle and the field disappears. The Higgs field does not possess this characteristic; instead, it exists regardless of the presence of a particle, with some hypothetical Higgs charge. In the language of particle physics, the Higgs field possesses a 'non-zero vacuum expectation value'.

Through interactions with Higgs force carriers (Higgs bosons), certain particles – such as quarks and electrons – gain mass, whilst others – such as photons – do not. If Higgs bosons do exist they will be rather massive particles, and will not be detectable by present particle accelerators. Upgrades at CERN, however, will allow the search for the Higgs boson to truly begin by the middle of the first decade of the twenty-first century. This is when the Large Hadron Collider (LHC) (see Figure 3.6) is scheduled to come on-line. Even if the Higgs boson is a figment of the imagination, physicists are hopeful that whatever generates mass in the Universe will be elucidated by the LHC.

3.5   UNIFICATION THEORIES

Throughout human history, what were once thought to be totally different forces have been shown to be different manifestations of the same force. Newton, for example, showed that the motions of the planets could be explained by the same force that caused the proverbial apple to fall to the Earth: gravity. Outwardly they were two totally different phenomena, but in reality they were manifestations of the same underlying principle. Earlier in this chapter we saw how electricity, magnetism and light were unified. Einstein showed an interconnection between space and time in special relativity, and then linked this work to gravity via his general theory of relativity.

The process of unifying the forces continues today, and particle accelerators have been the tool for taking the latest step by confirming a particularly remarkable theory. A group of physicists had proposed that electromagnetism and the weak nuclear force would act in exactly the same fashion at sufficiently high energies. Confidence in the Salem, Weinberg and Glashow theory was boosted enormously by the particle accelerator confirmations of their theoretical predictions. They had postulated that at sufficiently high energy it would become impossible to recognise the effects of the electromagnetic force apart from those of the weak nuclear force. Now electromagnetism and the weak nuclear force are forever linked as the

Figure 3.6. The superconducting magnetic test string for the Large Hadron Collider (LHC) is seen here in the assembly hall where it is being tested. (Photograph reproduced courtesy of CERN Geneva.)

electroweak force. This success has led physicists and cosmologists to believe that the strong nuclear force could also be unified with the electroweak force. It is postulated to take place at even higher energies, and is explained by a set of ideas known as the grand unified theories (GUT). This is a set of theories which sets out to unify all the fundamental forces except gravity. A number of GUT exist, but as yet none have been proven. GUT predict that protons should decay.

Some physicists and cosmologists believe that, following a successful GUT, gravity will be unified, and every interaction in the Universe will then be describable in terms of a single fundamental force of nature (see Figure 3.7). This level of unification, even if it exists, is a long way in the future. A quantum theory of gravity is required first.

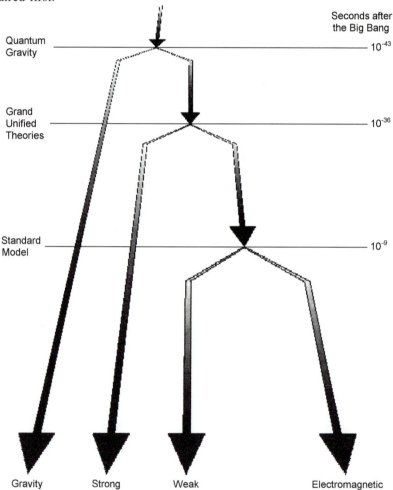

Figure 3.7. Unification of forces. Central to most cosmological ideas is the notion that, as the temperature of the Universe was higher at earlier and earlier times in history, so the distinction between the separate forces of nature was less and less apparent.

4

Observational cosmology

Cosmology is, by its very nature, an observational science. It is passive in that we must wait for events to happen in the Universe which we can then watch and study. Not for the practising cosmologist is the relative ease of making a Universe in the laboratory, although simulations of the early Universe are partially possible in particle accelerators. Luckily (as pointed out in Chapter 1) astronomers can rely on the experimental science of physics. Every piece of cosmological insight gained is achieved by painstakingly modelling observed celestial characteristics, using the physics gleaned from laboratory experiments.

The first fundamental observation which can be made about the Universe as a whole is that it contains matter. The radiation content is expected (as we shall see), but the fact that solid lumps of matter abound is rather a shock! The reason for the surprise is based on the results from experiments currently being conducted in the particle accelerators, (mentioned in Chapter 3). In these experiments, the interchangeability of mass and energy is explored. Energy, carried by photons, can be changed into mass under certain conditions: for example, if a photon closely approaches a heavy atomic nucleus. When this happens, two particles are produced – one of matter, and the other its antimatter counterpart. This is necessary to conserve charge and other quantities. For the purposes of this example, imagine that the two particles created are an electron and its counterpart, a positron. Eventually – and usually sooner rather than later – the positron comes into contact with the electron (or another which is just like it) and they annihilate each other, returning their energy in the form of photons (see Figure 4.1). If this process holds true universally, then for every particle of created matter there should be an antimatter equivalent. Eventually, mutual annihilation events will return all the mass energy as photons, leaving nothing from which to make stars, planets, you or me. Could it be that the antimatter has somehow been segregated from the matter? This is a rather unsatisfactory explanation, since there is no observational evidence to support it. It is also unsatisfactory from the point of view of the cosmological principle (discussed later in this chapter). Some way has to be devised, within the known laws of physics, which enables the ratio of matter particles to antimatter particles to be greater than 1. Current approximations based upon the ratio of photons to matter particles

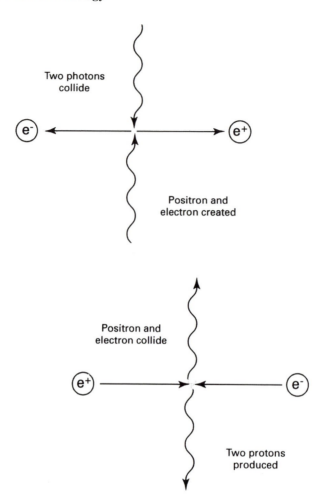

Figure 4.1. Pair production is the creation of a matter particle and its antimatter counterpart as a result of an interaction between two photons.

suggest that the asymmetry of matter to antimatter is probably only one part in a billion (1,000 million)! Very small indeed; but as insignificant as one in a billion may sound, it has led to a profoundly different Universe than one filled with radiation alone.

4.1   LOOK-BACK TIME

At first, the very idea of cosmology – observing the Universe as it exists now, so that we can tell how it all began – seems preposterous! It would be like taking a snapshot of a person, and telling his or her entire life story from that one single photograph!

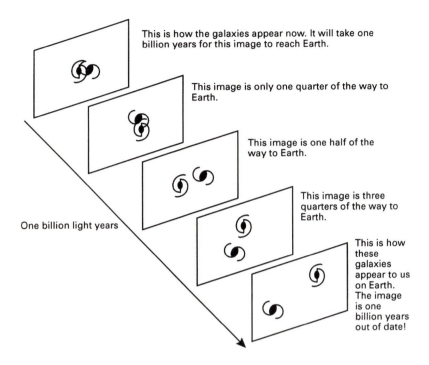

This is how the galaxies appear now. It will take one billion years for this image to reach Earth.

This image is only one quarter of the way to Earth.

This image is one half of the way to Earth.

This image is three quarters of the way to Earth.

This is how these galaxies appear to us on Earth. The image is one billion years out of date!

One billion light years

Figure 4.2. The finite speed of light causes the phenomenon of look-back time. The deeper astronomers look into space, the more ancient are the objects they observe.

Yet the aim of this chapter is to explain why cosmologists feel justified in doing this. A key element which requires immediate introduction is that light and all other forms of electromagnetic radiation can travel only at a finite speed.

As discussed in Chapter 2, the speed of light in a vacuum is unsurpassable, with a value of 3×10^8 m/s. On terrestrial scales, this means that events are communicated almost instantaneously. Over something as large as the Universe, however, events can take a long time to propagate; so long, in fact, that the more distant the object, the further back in time we are viewing it. To quantify this characteristic of the Universe, astronomers have developed a distance measurement system known as the light year. This is the distance travelled by light in a single year, and is equivalent to 9.4607×10^{12} km. If two stars are separated by a distance of one light year, then events which take place on one stellar surface are only apparent to the other after a period of one year, during which time the light released by the event has travelled the 9.4607×10^{12} km between the two objects. Closer objects will be able to observe the event earlier than can more distant objects. Thus, a galaxy which is a billion light years away will appear to us as it did a billion years ago. This phenomenon is known as look-back time (see Figure 4.2).

The information that an event has taken place propagates outwards in a spherical volume around the event, with a radius that increases with the speed of light. The

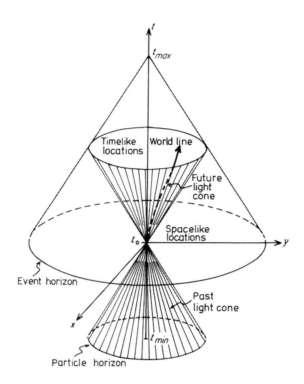

Figure 4.3. Light cones show the propagation of light from an event. If one event is situated within the light cone of the other, then they are causally connected. If one event is situated beyond the light cone of the other, then the two events are not yet aware of each other's existence. (Adapted from Roos, M., *Introduction to Cosmology*, Wiley, 1994.)

'surface' of the sphere is known as the event's horizon, but this should not be confused with the more common concept of an event horizon around a black hole, which is distinctly different. Over small volumes of space, an event which took place at time t will have a present-day (time = t_0) event horizon, r_e, of

$$r_e = c(t_0 - t_1) \tag{4.1}$$

Over large distances of space, relativistic effects will destroy the simplicity of this relationship. Diagrammatically, this behaviour is represented by a light cone (see Figure 4.3). Light cones are drawn on Minkowski spacetime diagrams, named after the physicist who first proposed the concept of spacetime. Anything which exists in the Universe can be plotted on a spacetime diagram, and will follow paths known as world lines. For unaccelerated travel, world lines are straight, but acceleration causes them to become curves. The gradient of a world line is an indication of the speed of the object. For example, a stationary object will have a world line gradient of either zero or infinity, depending upon the orientation of the time axis. Light rays, being the fastest thing in the Universe, follow paths, called null geodesics, through the spacetime continuum, and have a set gradient for their world lines. Objects within each other's light cones are causally connected; that is, they have been able to exchange photons, and hence can influence each other during their respective lifetimes. The two objects are then said to have a time-like separation. Objects not within one another's light cones are said to possess space-like separation.

If the light cone of the observable Universe is considered, then it becomes obvious that there could very well be objects which we do not know exist, because light from them has not yet reached us. Since the light cone of the Universe is defined by its age, any object which has a space-like separation from us cannot possibly have influenced our evolution. These unknowable regions are termed domains, and could conceivably have totally different laws of physics. The limit of our observable Universe is again known as an event horizon. Unfortunately we cannot observe it, because the cosmic microwave background blinds our sight – as we shall soon discover.

4.2   OLBERS' PARADOX

Olbers' Paradox is a fundamental observation which can be made with nothing more than the naked eye. Ask anyone to describe the night sky, and the chances are that one of the first things they will tell you is that it is dark.

Kepler (whose laws of planetary motion we encountered in Chapter 1) pointed out in 1610 that if the Universe is infinite with stars scattered uniformly throughout it, why is it dark at night? After all, in whatever direction one cares to look, our line of sight will, sooner or later, come to rest on the surface of a star. Although light suffers from an inverse square reduction in its intensity based upon its distance, the further one looks into space, the more stars will appear in our line of sight, compensating for the dimming. This curious conundrum was popularised over a century later by Heinrich Olbers, and now bears his name (see Figure 4.4).

The most obvious solution to this problem is that stars are not distributed uniformly through space; instead, they form galaxies. This provides only temporary respite, and the problem reappears with galaxies because, as we shall soon see, on the largest scale galaxies and clusters of galaxies can be thought of as being spread uniformly throughout space.

One solution to the problem, which provides a first fundamental insight into the nature of the Universe, derives from a consideration of look-back time. Imagine our Galaxy and another more distant galaxy to be plotted on a spacetime diagram. We become aware of the other's existence (and *vice versa*) only when their light cone crosses our world line. If the Universe is not infinitely old, then there must be distant galaxies with light cones not yet in contact with us.

In order to resolve Olbers' Paradox we have been forced to assume that the Universe may not be infinite in age; and in doing so we have taken our first step towards the Big Bang, because it implies that at some time in the past the Universe must have been created. Another complementary resolution also exists, as will be explained in the subsequent sections.

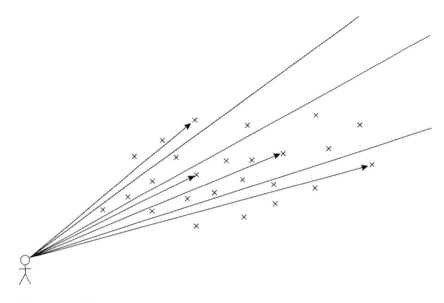

Figure 4.4. Olbers' Paradox asks why the sky is dark at night. If the Universe were infinite in extent, then our line of sight would eventually meet with the surface of a bright star. The reason the night sky is dark, therefore, is because the Universe is not infinitely large, and because redshift renders distant objects more and more faint.

4.3   THE DOPPLER EFFECT

The galaxies in the night sky appear static until they are studied spectroscopically. Astronomer Vesto Slipher discovered that the spectra of galaxies display spectral absorption lines which have been shifted from their measured laboratory wavelengths. This can be explained as the optical equivalent of the Doppler effect (see Figure 4.5). We are all familiar with the way in which bells and sirens change pitch as they pass us. Instead of thinking of cars and trains which emit sound, let us transpose the idea to starships and light rays, and gain an entry route into thinking about how the Doppler effect changes rays of light.

Imagine two starships heading out into deep space. Initially, they are both travelling with the same velocity, although starship A is in front of starship B. Atop each of the spacecraft is a flashing light, each of which regularly pulses on and off with a frequency, f. Someone on starship B can see both the flashing light on his ship and the light on starship A. If this person measures the frequency of both lights it will be found that they are the same. If starship A suddenly increases to another velocity and the frequencies of the two flashing lights are again compared, it will be noticed that f_A is smaller than f, but that f_B remains the same (equal to f). Thus, a fundamental change in the way an observer on starship B perceives starship A has taken place, simply because there is now a difference in velocity between the two vessels. This concept is at the heart of Einstein's special relativity, and is a direct

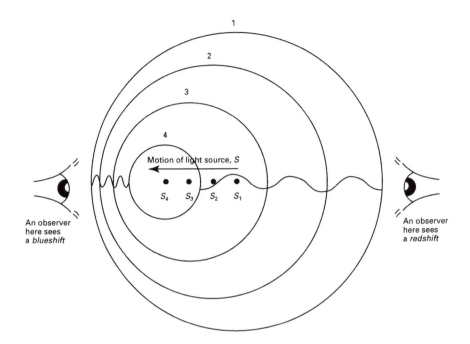

Figure 4.5. Motion of either the source or the observer causes a Doppler effect to be imparted to radiation. (Adapted from Kaufmann, W.J., *Universe*, W.H. Freeman, 1987.)

consequence of light possessing a finite, invariant velocity through space. But does this mean that the flashing light has actually slowed down, or does it simply mean that starship B has perceived it to slow down? This is where the concept of relativity comes into play. The flashing light on starship A appears to slow down only to observers in motion away from it. It does not matter whether the observer is moving away, whether starship A is moving away, or whether both are moving away from each other. All that matters is that they are in relative motion away from one another (that is, the distance between them is increasing with time). The exact amount by which the flashing light slows down is then dependent upon how fast they are separating. Since it does not matter which starship is moving, from an observer on starship A's point of view, it is the flashing light on starship B which has slowed down, not the one on starship A.

The concept is easy to grasp through some elementary mathematics. If two observers, one on each vessel, agree to time the flashes of starship A's light, then observer A records a time, t_{A1}, as soon as the light flashes on (see Figure 4.6). The observer on starship B begins timing after the light has crossed the distance, d, between the two ships. Thus, the time recorded by observer B is

$$t_{B1} = t_{A1} + \frac{d}{c} \tag{4.2}$$

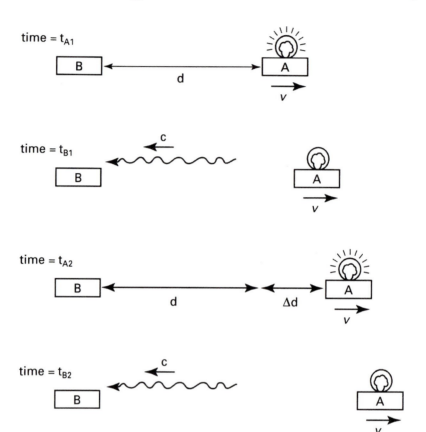

Figure 4.6. Starships A and B are separated by a distance, d, which is getting larger all the time because A is moving away from B with a velocity, v. If an observer on A and an observer on B compares the length of time between flashes of a light on A, it will be noticed that the time recorded by the observer on B is longer than that recorded by the observer on A. (See main text for a full explanation of this phenomenon and the underlying mathematics.)

When observer A sees the second flash a second time, t_{A2}, is recorded. Thus, the time between flashes, as timed by observer A, is

$$\tau = t_{A2} - t_{A1} \tag{4.3}$$

τ is known as the proper time because it is the time of the flashes as observed from the light's own frame of reference (that is, there is no relative motion between the light and the observer). The observer on starship B records a time of

$$t_{B2} = t_{A2} + \frac{(d + \Delta d)}{c} \tag{4.4}$$

This distance, Δd, is how far the starships have parted in the proper time, τ. From equations (4.2) and (4.4), the time interval, Δt, between flashes, as measured by observer B, is defined as

$$\Delta t = \tau + \frac{\Delta d}{c} \qquad (4.5)$$

Thus, the increase in the length of time between flashes is proportional to the increase in distance between the two ships, which in turn is proportional to the relative velocity between the two ships. If the distance between the ships decreases, then Δd will be negative, and $\Delta t < \tau$.

It is important to remember that this section has analysed what observers A and B see of the flashing light on starship A. If it were recast to analyse the appearance of the flashing light on starship B, it would be observer A, not observer B, who sees the increased time. This reinforces the fact that there are observational effects which have to be considered when we observe objects in motion relative to ourselves.

4.4   REDSHIFT

To understand the Doppler effect, imagine that the flashing light on starship A is replaced by a steady light source. Instead of an increase in the flash period, the time between electromagnetic wave crests will increase as the two ships become further apart. This will obviously increase the wavelength of the light and hence shift it towards the lower-energy end of the electromagnetic spectrum. In visible light, the lower-energy colour is red, and so this phenomenon is known as the redshift. If the distance between the two starships had been decreasing, the shift would have been to the blue (higher-energy) end of the electromagnetic spectrum. The redshift, z, is quantified by equating the rest wavelength, λ, with the observed wavelength, λ_0:

$$z = \frac{\lambda_0 - \lambda}{\lambda} = \frac{\lambda_0}{\lambda} - 1 = \frac{\Delta\lambda}{\lambda} \qquad (4.6)$$

Equation (4.5) showed that the increase in flash period was proportional to the distance travelled by the starships between flashes. This distance is obviously proportional to the radial velocity, v_r, of the starship. If we now equate the flash period with the time between wave crests, the increase in the flash period becomes analogous to the redshift, and we can rewrite equation (4.5) as:

$$z = \frac{v_r}{c} \qquad (4.7)$$

Earlier in this chapter it was stated that equation (4.1) required modification to account for relativistic effects. This equation holds true only for certain ranges of values: specifically, $|v| < 0.1c$ (see Figure 4.7). Above this velocity range, relativistic modifications must be introduced in the following way.

Imagine two frames of reference, S and S', moving away from one another in the x

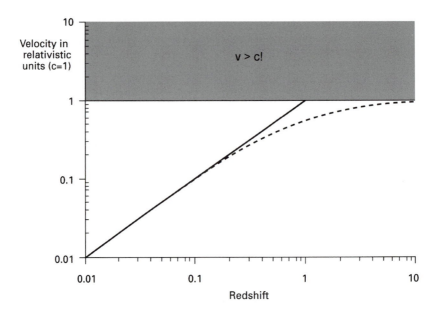

Figure 4.7. The linear redshift approximation (solid line) can be deduced only when velocities are below one-tenth of the speed of light. Above this figure, the full relativistic formula (dashed line) must be used.

direction. If we are observing from S, then S′ appears to be moving with velocity v. The time ordinate is set at zero when S coincides with S′. An external event occurs at some further time, t, and is measured by ourselves and an observer at S′. In order to compare our measurements to verify that they are the same, some way of transforming our co-ordinates from S to S′ must be found. Classical physics uses a set of equations known as the Galilean transformation:

$$x' = x - vt$$
$$y' = y$$
$$z' = z$$
$$t' = t \tag{4.8}$$

This implicitly assumes that light travels instantaneously between the event and each of the observers. In normal circumstances this is quite a reasonable assumption; when an electric light is turned on in a room, everyone, regardless of their position within the room, sees it simultaneously. When light travels over large distances, however, time delays between observers at different positions in different frames of reference will be noticed. This is because light does, indeed, have a finite velocity, although it is extremely fast – 3×10^8 m/s. As soon as the relative velocities of sources and observers grow to more than 10% of the speed of light, a totally new set of transformations must be incorporated, because different observers in different frames of reference may not record the event at the same time. It also takes into account the observational verification that light appears to have the same velocity regardless of the velocity of the observer. The new transformations were originally derived empirically by Lorentz to explain the failure of the Michelson–Morley experiment to detect the ether. Later, in his special theory of relativity, Einstein established them again, but from a foundation of sound theoretical arguments. Nevertheless, they retain their original name of the Lorentz transformations:

$$x' = \frac{(x - vt)}{\sqrt{\left(1 - \frac{v^2}{c^2}\right)}}$$

$$y' = y$$
$$z' = z$$

$$t' = \frac{\left(t - \frac{vx}{c^2}\right)}{\sqrt{\left(1 - \frac{v^2}{c^2}\right)}} \tag{4.9}$$

It is implicit in this transformation that time is affected by velocity. This is a phenomenon which has been termed 'time dilation'. Since the wavelength is related to the time between successive emissions of a wave crest, it becomes obvious that time dilation must also cause a redshift. Instead of being the radial component of velocity, time dilation depends upon the overall velocity of the source or observer, v. When this is taken into account, equation (4.6) becomes

$$z = \frac{\left(1 + \frac{v_r}{c}\right)}{\sqrt{\left(1 - \frac{v^2}{c^2}\right)}} - 1 \tag{4.10}$$

If the velocity is entirely radial, $v = v_r$, the equation simplifies to

$$z = \sqrt{\left(\frac{1 + \frac{v}{c}}{1 - \frac{v}{c}}\right)} - 1 \tag{4.11}$$

This can be rearranged to represent the velocity of separation between the source and observer based upon a measurement of the source's redshift:

$$\frac{v}{c} = \frac{(1 + z)^2 - 1}{(1 + z)^2 + 1} \tag{4.12}$$

Using this equation, redshifts larger than 1 no longer yield the non-physical solution that the distance between the source and the observer is increasing at a velocity greater than the speed of light.

4.5   THE HUBBLE FLOW

Vesto Slipher's work showed that in the majority of galaxies studied by him the wavelength of the spectral lines shifted to the red end of the spectrum. Only in a few rare cases did the lines shift towards higher energies.

Slipher's observations were extended and finally explained by astronomer Edwin Hubble. As well as developing his morphological classification scheme for galaxies, he was also dedicated to determining the distances of galaxies, and in 1924 finally showed that the Andromeda galaxy and M33 are too far away to be contained within our Milky Way. He then began to determine the distances of other galaxies and, in the process, began to realise that a trend in the data was becoming evident. The further away the galaxy, the greater the redshift seemed to be. With the help of fellow Mount Wilson astronomer, Milton Humason, Hubble investigated this dependence and found it to be a linear proportionality. With the exception of a few nearby galaxies which are gravitationally bound to the Milky Way, every other galaxy seemed to be displaying a redshift, indicating that the distance between it and the Milky Way was increasing. The linear nature of the proportionality indicated that if two galaxies existed – one at twice the distance of the other – the nearer galaxy would be moving away from us with only half the speed of the further one. This is a property known as uniform expansion. The expansion of the Universe around us, which causes the galaxies to move away from us, is known as the Hubble flow. It is a cornerstone of the Big Bang theory.

But is it realistic to think of galaxies rushing with these velocities through space? Hubble referred to them as apparent velocities, and it is now known that the galaxies do not actually move through space at their calculated recessional velocity. Instead, it is the spacetime continuum between the Milky Way and each galaxy that expands. The galaxies themselves are more or less stationary, although they may possess some orbital motion around neighbouring galaxies. And because it is the spacetime continuum that is expanding, the special relativistic equation (4.12) is no longer needed; the recessional velocity of a galaxy can be calculated by the simple equation (4.7). Since nothing is actually moving through space, the velocity can be greater than the speed of light, and no physical laws are violated. The analogy is that dots painted on a balloon move away from each other when the balloon is inflated, but they do not actually move across the surface. In the Universe, galaxies are the dots on the cosmic balloon – the spacetime continuum.

Hubble and Humason's redshift–distance law also resolves Olbers' Paradox by consistently reducing the energy of the light coming from distant galaxies. This makes them appear ever fainter, and hence more difficult to observe.

In the first half of the twentieth century the Russian astronomer Aleksandr Friedmann and the Belgian priest and cosmologist George Lemaître each, independently, predicted the Hubble flow two years before its observational verification. Lemaître went on to postulate that the Universe must have begun as a compacted primeval atom which then exploded and scattered matter throughout space. He based this idea on the fact that since the galaxies were all moving away from one another, they must once have been very close together. With this concept,

Lemaître planted the idea of a creation event in the mind of the scientific community.

4.6   THE DISTRIBUTION OF GALAXIES THROUGHOUT SPACE

As well as his other pioneering work on galaxies, Hubble sought to understand how galaxies are distributed throughout space. Indeed, this type of work still occupies a huge amount of time spent by the modern observational cosmologist. Surveys of ever-increasing numbers of galaxies are attempting to elucidate the way in which these collections of stars are distributed throughout space. These surveys all rely upon measurements of redshifts, which can then be converted into distances.

As explained in Chapter 1, it has become obvious that galaxies aggregate in groups and clusters. These clusters are then grouped into superclusters, which are strewn throughout space surrounding huge voids. Pencil beam surveys probe deeply into the Universe, revealing the distribution of galaxies along very specific lines of sight. The surveys of Broadhurst and Koo have clearly shown that galaxies are not distributed randomly, but occur periodically throughout space. The regions containing no galaxies correspond to the voids, whilst the clumping of galaxies simply signifies the position of a supercluster on the void edges.

4.7   THE ISOTROPY AND HOMOGENEITY OF THE UNIVERSE

If the Universe around us were to be divided into cubes with dimensions of 10 million light years on each side, and the density of galaxies within them studied, some cubes would contain more galaxies than would others. The differences would be so noticeable that, whereas some cubes would contain entire groups and clusters of galaxies, other would contain nothing! If we were to increase the dimensions of the cube edges by a factor of 10 to 100 million light years, the situation would become slightly less radical, as every cube would contain galaxies. In fact, every cube would contain clusters of galaxies, because this is on the scale of the superclusters. However, the average density of galaxies would still be variable, because the Universe is punctuated by large voids which are empty of galaxies. In order to make each cube contain roughly the same quantity of galaxies, its linear dimensions must be increased beyond 100 million light years. The precise amount by which the scale must be increased is still a matter of some debate but, based upon the Las Campanas Deep Redshift Survey, which has collected over 12,500 galaxy redshifts, the scale is in the order of 300–600 million light years.

If the entire contents of the Universe were to be taken apart and spread throughout space as a uniformly tenuous gas – a concept known as the cosmic substratum – each cubic metre of space would contain only one nucleon! The density of this cosmic substratum would be approximately 10^{27} times less than the density of the Earth's atmosphere at sea level. This has led cosmologists to assume that the matter content of the Universe can be thought of as a perfect fluid; that is, one which does not resist changing its shape under external influence.

Thus, on a scale greater than 100 million light years, the Universe can be thought of as homogeneous – the same at all points throughout space. This result is reached regardless of the direction in which the observer looks. A cosmologist looking due north would reach the same conclusion as an astronomer looking due south. This means that the Universe looks the same in all directions, and so is said to be isotropic. An isotropic Universe adheres to the Copernican principle, which states that the Universe has no preferred location within it. No matter where we are in space, we will see a totally representative part of the Universe. Nowhere will we be able to see a different class of star or galaxy, because they are all scattered homogeneously throughout the cosmos, and so no location takes precedence over any other. An isotropic Universe can have neither a centre nor an edge, which would violate the isotropy. The view from an edge would be very different than a view from the centre (see Figure 4.8).

4.8   THE COSMOLOGICAL PRINCIPLE

The observational conclusion that the Universe is both isotropic and homogeneous has been raised to the level of a guiding principle in our study of the Universe. It is known as the cosmological principle, from which we can actually deduce the observationally proven Hubble flow in the following way.

Imagine two observers, A and B, situated in different galaxies in the Universe (see Figure 4.9). Observer A measures the mean density of galaxies, ρ, in two different directions along radius vectors $\mathbf{r}$ and $\mathbf{r}'$. The cosmological principle dictates that as long as observer A does this at the same cosmic epoch:

$$\rho_B(\mathbf{r}) = \rho_A(\mathbf{r}') \tag{4.13}$$

Similarly, if observer B measures the mean density of galaxies in direction $\mathbf{r}'$ in the same epoch as observer A:

$$\rho_B(\mathbf{r}') = \rho_B(\mathbf{r}) = \rho_A(\mathbf{r}) = \rho_A(\mathbf{r}') \tag{4.14}$$

Thus, if A and B turn their attention to another measurable cosmic quantity – such as the velocity distribution, v, of the galaxies they can observe – the cosmological principle states that

$$v_B(\mathbf{r}') = v_A(\mathbf{r}') \tag{4.15}$$

If observer B has a relative velocity of $v_A(\mathbf{r}'')$ as measured by observer A, providing that the velocities are below 10% of the speed of light,

$$v_A(\mathbf{r}') = v_A(\mathbf{r} - \mathbf{r}'') = v_A(\mathbf{r}) - v_A(\mathbf{r}'') \tag{4.16}$$

Equation (4.16) holds true, however, only if

$$v_A(\mathbf{r},t) = f(t)\mathbf{r} \tag{4.17}$$

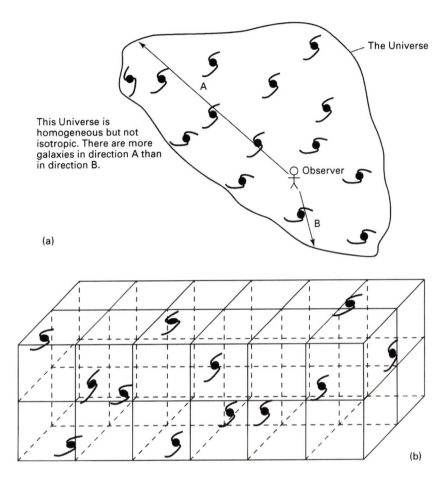

This Universe is
homogeneous but not
isotropic. There are more
galaxies in direction A than
in direction B.

The Universe

A

○ Observer

B

(a)

(b)

Figure 4.8. The isotropy of the Universe (a) implies that it has neither a centre nor an edge, and the homogeneity (b) allows astronomers to make sweeping statements about the cosmos because we live in a representative part of it.

Here we have explicitly stated the time dependence of the equations by inserting the variable, t, into the functions. For the present epoch, time is a constant, and so equation (4.10) reduces to the Hubble equation (1.4).

The function f(t) can yield either a positive, negative or zero value for any given time. A negative value would correspond to a collapsing Universe, whereas a positive value would indicate that the Universe is expanding. The third possibility, $f(t)=0$, can be discounted if we resort to the following analogy.

Imagine spaceships blasting off from the Earth. Any vessel which does not attain the Earth's escape velocity (~ 11 km/s) will be dragged back by the gravity of our world. This is said to be a closed gravitational system, and corresponds to $f(t)<0$. Any craft which increases its velocity beyond this will escape the gravitational influence of the Earth, creating an open gravitational system, ($f(t)>0$). In the special

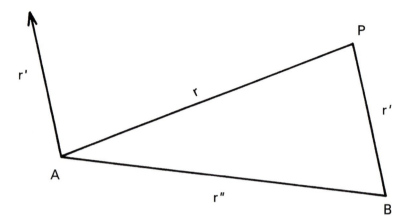

Figure 4.9. The cosmological principle states that the Universe is both isotropic and homogeneous; that is, we see an entirely representative part of the Universe. This concepts allows us to predict the Hubble flow. (Adapted from Roos, M., *Introduction to Cosmology*, Wiley, 1994.)

case of the spacecraft which reaches the exact escape velocity, it enters into orbit around the Earth. This is known as the static solution, because $f(t)=0$.

If the static solution is applied to the Universe at large, it implies that there is a preferred location around which bodies can orbit. This is expressly forbidden by both the Copernican principle and the cosmological principle. Thus, by a simple consideration of what we know about gravitational systems, and by applying the cosmological principle, we have shown that the Universe is expected to be either contracting or expanding.

The other reason why the static solution for the Universe is highly improbable stems from a consideration of gravity. In formulating his now famous law of gravity, Newton expressed the view that the Universe might be collapsing under the force of its own gravity. In other words, he wondered how the cosmos could possibly be static when it contains massive bodies which would naturally pull everything together. He failed to pursue this line of reasoning, however, and the static Universe model persisted until it was superseded by the work of Slipher, Hubble and Humason.

4.9   THE HELIUM PROBLEM

Having looked at kinematic constraint, we shall now turn our attention to a compositional one. The use of spectroscopy has allowed astronomers to deduce the chemical abundance of elements in the Universe at large. This work has led to the intriguing problem of how the Universe came to be populated by the chemical elements in the relative abundances found by spectroscopists.

Estimates have shown that, if one were to sample 100 atoms from an average cosmological environment (not from a planet such as Earth), the breakdown would reveal that about 93 of the atoms would be hydrogen, whilst the other seven would be helium. Occasionally, one or possibly two of the atoms encountered would be from a heavier element.

Translating this into the mass of the chemical elements which make up the Universe: 75% of the cosmic mass would be hydrogen, 23% would be helium, and 2% would be everything else! This is far too much helium for it all to have been produced inside stars. If it had been, the mass of heavier elements would be far larger than the 2% we observe today.

It is now known that stars produce their energy by the nuclear fusion of hydrogen into helium. Hydrogen is the simplest element of nature, consisting of just a single electron in orbit around a single proton. The fusion process builds elements by adding particles to the atomic nucleus. Thus, an analysis of the chemical composition of a star should reveal the age of the star. Stars which contain a relatively large amount of heavier elements are known as Population I stars, whilst earlier stars are termed Population II. Some astronomers also believe that an even earlier population of stars can be found. These Population III stars, however, have yet to be discovered.

Analysis of the oldest stars available to us presents an interesting problem. Whilst the levels of heavier elements are acceptable, the amount of helium contained in these stars is far greater than could possibly have been created in their violent hearts. It is as if the Universe had a predisposition to helium which was quite out of proportion when compared with the other elements.

This excess has become known as the helium problem, and is one of the fundamental observational properties which must be taken into account by any viable cosmological theory. The first steps towards an explanation were taken by George Gamow in the 1930s. He sought to understand how chemical elements had been formed in the quantities in which they were observed. At this stage in history, the role of nuclear fusion in stars was not yet understood. Gamow knew that the chemical identities of different elements were determined by the differing number of protons that populated their nuclei, and he also realised that by adding or subtracting neutrons and protons to a nucleus, artificial elements could be created. The problem was that, even if starting with a basic mixture of neutrons and protons, how could they be forced together and retained in that form? The answer, Gamow discovered, was to subject them to extremely high temperatures: billions of Kelvin. His next problem was to answer the question: where in the Universe can such a titanic inferno be found? Gamow realised that, at the time, the answer was 'nowhere', but, building on the work of Lemaître, he produced one of his greatest insights by likening the Universe to a bicycle pump. If air is compressed in a pump, the atoms and molecules in the air heat up. If the Universe were to be compressed in such a way – as would have been the case in the early Universe when everything was very much closer together – he reasoned that it must have been very, very hot – so hot, in fact, that it provided the temperatures necessary for nuclear fusion to take place.

Unfortunately, Gamow's theory worked well only for explaining the 23% of helium; the other, heavier elements could not be explained. As a result his theory was not as celebrated as it should have been, and the idea was only slowly accepted. We now know that almost all of the other elements are produced in the hearts of massive stars and returned to space in supernova explosions. Supernovae alone are incapable of synthesizing the abundance of helium in the Universe. In essence, Gamow had successfully solved the helium problem before it had ever really become a problem! In doing so, he was the first cosmologist to ever seriously (and quantitatively) consider the first moments after creation; and he also proved that the Big Bang was not only a dense phase in the Universe's existence, but also a very hot one! The concept of a hot Big Bang had been born.

4.10 THE COSMIC MICROWAVE BACKGROUND RADIATION

The concept of look-back time has allowed us to view the Universe at successively earlier and earlier times in the history of the Universe – usually referred to by the generic term, epochs – simply by looking further and further into space. If the Universe has always existed then, in principle, one should be able to observe further and further into space, seeing nothing but galaxies which look exactly the same as those around us. However, what we actually observe in the cosmos is that the types of celestial objects change, the further into space we look. This implies some sort of evolutionary effect, and perhaps indicates that the Universe is of finite age. If this is the case, then the question naturally arises: can we view the epoch of creation if we look far enough out into space? Unfortunately, the answer is 'no'. A fundamental barrier has been discovered to exist. This barrier is characterised by the emission of microwave radiation, which blinds us to earlier events and blocks our view of the early Universe.

The release of this background radiation signifies a fundamental change in the characteristics of our Universe, and was actually predicted by Gamow's colleagues Ralph Alpher and Robert Herman in 1948. It is further proof of a hot Big Bang, because any hot object emits thermal radiation (as described in Chapter 2) and, if the constituents of the Universe were at billions of Kelvin, they must have given out copious quantities of radiation in the gamma-ray region of the spectrum. Alpher and Herman predicted that the radiation would now look like a thermal source at a temperature of about 5 K.

Microwave radiation was discovered, serendipitously, in 1964 by Arno Penzias and Robert Wilson (see Figure 4.10). Whilst preparing the radio horn at Bell Telephone Laboratories, New Jersey, for use in radio astronomy, the pair of researchers could not eradicate a stubbornly persistent signal. This annoying 'hiss' was eventually shown to be the first observed point on the graph of the cosmic microwave background radiation's spectrum.

Penzias and Wilson's data point was actually a measure of the CMBR's intensity at a wavelength of 7.35 cm. Work was swiftly begun to measure the background intensity at other wavelengths, but the atmosphere absorbs quite strongly close to the theoretical maximum of the background radiation, and the experiments were severely hampered. Nevertheless, more and more data points were constantly being

Figure 4.10. Dr Arno A. Penzias and Dr Robert W. Wilson, of the Bell Telephone Laboratories, who in 1964 discovered the microwave background radiation whilst using the horn-shaped antenna partially shown in the background of this picture. The antenna was designed to relay telephone calls to communications satellites in Earth orbit. (Photograph reproduced courtesy of Bell Telephone Laboratories.)

added to the curve, and a thermal curve gradually began to take shape. Absolutely incontrovertible evidence that the CMBR was a black-body curve (see Figure 4.11) was obtained by the Cosmic Background Explorer (COBE), a satellite launched by NASA in 1989.

When COBE measured the microwave background radiation, it found it to be characterised by a temperature of 2.726 ± 0.005 K, which was very close to what had been predicted by Alpher and Herman. This temperature has allowed the number density of CMBR photons to be calculated. We can expect to find an average of 400 cosmic microwave background photons in every cubic centimetre of space – an overwhelming number of photons which totally dwarfs the number of photons produced by starlight or any other celestial process. Thus, the cosmic microwave background radiation is the dominant source of radiation in the Universe.

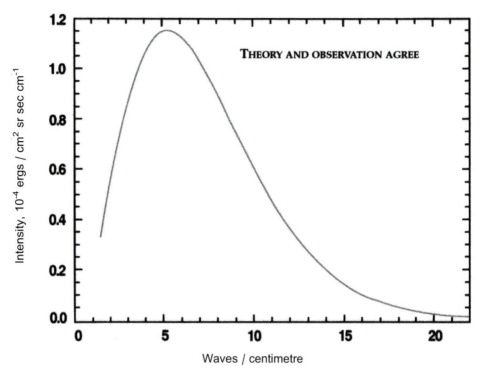

Figure 4.11. The spectrum of the cosmic microwave background radiation fits very precisely with a black-body curve of a thermal source at 2.7 K. This spectrum of the CMBR obtained by COBE provides firm evidence of the Big Bang. (Photograph reproduced courtesy of COBE Science Working Group, NASA, GSFC, NSSDC.)

In the discussion on the concept of the substratum (at the beginning of this chapter) it was stated that the average matter density of the Universe is just one nucleon per cubic metre. This leads to the ratio stated earlier: for every particle of matter there are approximately 1,000 million photons. Therefore even today the content of the Universe is totally dominated by the left-over radiation of the Big Bang. This does not mean, however, that the energy content of the Universe is dominated by the CMBR. As we saw in Chapter 2, the amount of energy contained by a photon is proportional to the frequency of the radiation it represents (equation (2.5)). Microwaves are, energetically speaking, rather weak. If the total energy bound up in the microwave background, e_r (using equation (2.5)) is compared with the energy bound up in matter, e_m (using equation (3.2)), then $e_m \gg e_r$. Hence, in the present-day Universe, matter contains the dominant form of energy. This is why the Universe is said to be matter-dominated. However, this has not always been the case.

If the cosmic expansion is traced back through time, the redshift which has been imparted to the CMBR is reversed, and hence each background radiation photon increases in energy. Eventually a time is reached when the dominant form of energy in the Universe is in the form of the background radiation. The approximate condition for this occurs when each photon contains 1,000 millionth of the energy bound up in a proton. The proton is used because it is the dominant, stable heavy particle in the Universe; electrons contain little mass and there are too few neutrons to disturb the approximate nature of our calculations. The time at which the photon energy density and matter energy density were equal corresponds to the dividing line between our present matter-dominated Universe and the early radiation-dominated Universe.

This dividing line occurred about 100 seconds after the Big Bang; and because it happened before 300,000 years – when, as we shall see, matter and radiation stopped interacting on a large scale – there is an important consequence for the radiation. As we shall see in later chapters, the fact that matter came to dominance before it stopped interacting with the CMBR means that it left an imprint, showing us the primordial density distribution of matter in our Universe. These tiny fluctuations in the temperature of the background radiation were detected by the COBE satellite, and provide an insight into how galaxies form.

The observations presented in this chapter are those which have led directly to the Big Bang theory of the Universe's creation. They are not the beginning and end of observational cosmology, as we shall see in later chapters. The density of matter in the Universe, the evolution of galaxies and the search for dark matter are all important observations which, when analysed within the Big Bang framework, lead to some surprising and sometimes profound conclusions.

Table 4.1. The cosmic abundance of elements.

Element	Number of atoms, relative to silicon
Hydrogen	3.18×10^4
Helium	2.21×10^3
Oxygen	22.1
Carbon	11.8
Nitrogen	3.64
Neon	3.44
Magnesium	1.06
Silicon	1
Aluminium	8.5×10^{-1}
Iron	8.3×10^{-1}
Sulphur	5×10^{-1}
Calcium	7.2×10^{-2}
Sodium	6×10^{-2}
Nickel	4.8×10^{-2}

5

The Big Bang

Analysis of the observations discussed in the previous chapter has led cosmologists to the currently accepted model of the Universe's creation. It is known as the Big Bang – a name which was actually coined as a term of derision by Fred Hoyle. He and his colleagues Hermann Bondi and Thomas Gold proposed a different interpretation of the results, known as the Steady State model, but the detection of the microwave background radiation ultimately proved them wrong.

George Gamow was the first cosmologist who was sufficiently audacious to believe what his equations told him about the early Universe. The Big Bang is, indeed, our best model yet of how the Universe began, but it is important to remember that it is not yet proven! This chapter will provide a largely qualitative account of the Big Bang. It is here that we encounter the application of quantum theory to the Universe at large.

If the Universe is expanding – which our observational evidence would lead us to conclude – then, by extrapolating that expansion back through time, we arrive at a singular point in history at which the Universe was compacted into an infinitely dense region. This is what is known as the Big Bang. As we shall see, theoretical physics cannot yet take us back to the point of the Big Bang, but perhaps it will when quantum gravity theories are perfected. It may even show that the initial Big Bang point can be avoided altogether, and demonstrate that the Universe had remained in a quiescent state for an indefinite period of time before something sparked the current headlong expansion.

Discussion of an initial point of creation naturally leads most people to wonder what there was before the Universe existed. The answer is that space and time came into existence at the moment of the Big Bang. This being the case, it seems a little trivial to state that space and time did not exist before the Big Bang. It is an important statement, however, because it answers the question for us. If time was created at the instant of the Big Bang, the concept of 'before the Big Bang' ceases to have any meaning, because time itself did not exist. By a similar line of argument it can be reasoned that the fact that space expands does not imply that it has to expand into anything.

5.1   CHARACTERISING THE EXPANSION

The evolution of the Universe is driven by its expansion. The interlinked physical properties of temperature and energy allow cosmologists to calculate the conditions in which the early Universe is thought to have existed. In order to place our discussion of that evolution in context, we must first consider how the expansion of the Universe alters its temperature and energy.

In physics, an adiabatic process is defined to be one in which heat neither enters nor leaves the system under investigation. Therefore, if the system being studied happens to be the Universe as a whole, it must be adiabatic. Thus, the Universe expands adiabatically.

The concept of the cosmic substratum (introduced in Chapter 4) postulates that on the largest scales, the Universe can be thought of as an idealised, smooth fluid. At earlier and earlier times, this analogy becomes applicable over smaller and smaller distances. Before the age of 300,000 years the Universe is thought to have been an excellent example of an ideal fluid. This fluid is non-viscous, and all measurements are made from a co-moving frame of reference, which means that it is at rest within the substratum (see Figure 5.1). The constituents of the Universe – the particles of matter and the photons of energy – are all in thermal equilibrium during the early stages of the Universe.

At any point in the evolution of the Universe, its temperature, T, can be specified with reference to its energy, E, by the simple law:

$$E = kT \tag{5.1}$$

where k is the Boltzmann constant.

Thermodynamics defines a property known as entropy, which is a measure of the ease with which a system uses its internal energy to do external work. For an adiabatic process, the entropy of the system does not change and so the second law of thermodynamics requires that

$$dE = -pdV \tag{5.2}$$

where dE is the change in energy induced by a change in the system's volume, dV, assuming that the pressure, p, has remained constant. In a non-viscous fluid, the pressure is defined to be

$$p = \frac{1}{3} \varepsilon \tag{5.3}$$

where ε is the energy density.

For the moment we shall concentrate upon the energy density of the radiation content of the Universe, ε_r, because it drives the thermal evolution of the early Universe. As the Universe expands, both its volume and its energy density must change. If the linear size of the Universe is characterised by the function R(t) – which has the dimensions of distance and increases with time, t – the volume changes in proportion to the cube of this function because

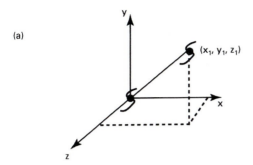

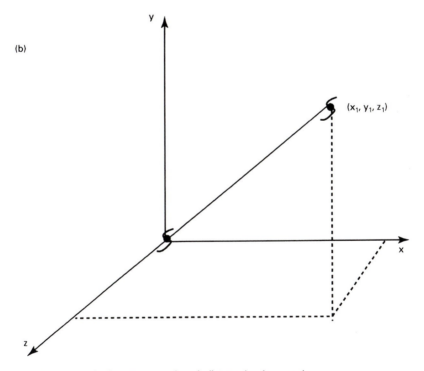

As the Universe expands so does the frame of reference:-

Co-ordinates remain the same even though distance has increased.

Figure 5.1. In a co-moving frame of reference, the co-ordinates of an object do not change with the expansion of the Universe.

$$V = \frac{4\pi R(t)^3}{3} \tag{5.4}$$

If we choose a small enough volume element to study, then the energy density is defined very simply, because we assume that only one photon is found in each element:

$$\varepsilon_r = \frac{h\nu}{V} = \frac{hc}{V\lambda} \tag{5.5}$$

Thus, ε_r changes in proportion to $R(t)^{-4}$ because it is not only dependent upon the expansion of the volume element but also upon the redshift of the photon. Remembering (from Chapter 4) that as time passes the Universe expands, the redshift of the radiation is proportional to $R(t)$. Hence, when the volume and the energy density are equated, the volumes cancel out, and we are left with

$$E \propto R(t)^{-1} \tag{5.6}$$

When this is substituted into equation (5.1),

$$T \propto R(t)^{-1} \tag{5.7}$$

So as space expands, the temperature associated with the radiation within it decreases. As stated in Chapter 4, the cosmic microwave background is the dominant form of radiation in the Universe, so it is this which governs the temperature of the Universe. There are so many photons of background radiation that any cold object placed in space will be impacted by the photons and raised in temperature until it mirrors that of the CMBR.

Having stated that our physical model for the early Universe is that of an ideal gas, cosmologists have to keep stock of all the individual particles which make up that cosmic gas. Chapter 2 introduced the Planck distribution (equation (2.4)), which allowed us to calculate the number of photons within a specific frequency (energy) range. A similar type of equation can be formulated for particles travelling at the relativistic velocities. Instead of a frequency range, this distribution denotes the number of particles within a range of momenta $(p + dp)$. Momentum is used because it contains the variables of mass, m, and velocity, v, which are used to define the total energy of the particle, $E(p)$, (rest mass energy + kinetic energy):

$$n(p)dp = \frac{8\pi}{h^3} \frac{n_{spin}}{2} \frac{p^2 dp}{e^{E(p)/kT} \pm 1} \tag{5.8}$$

The only variable not yet mentioned in this equation is that of n_{spin}, which is a measure of the quantum property (spin) of the particle under investigation. The equation is used for bosons (particles which do not obey the Pauli exclusion principle) by setting the ± 1 condition to -1. Fermions – which do obey the exclusion principle – are catered for by setting the condition to $+1$.

The Planck distribution can be easily derived from this equation by remembering that photons are bosons with an n_{spin} of 2. The particle energy, $E(p)$, is replaced by

Table 5.1. Particles, rest energies and relativistic domains.

Particle		Rest mass energy		Relativistic domain
		(MeV)	(K)	(K)
γ	photon	0, 0		
ν_e	electron neutrino	$< 10\text{--}5$	$< 10^5$	$< 10^6$
ν_μ	muon neutrino	< 0.27	$< 3.1 \times 10^9$	$< 3.1 \times 10^{10}$
e^-	electron	0.511	5.93×10^9	5.93×10^{10}
ν_τ	tau neutrino	< 3.1	$< 3.6 \times 10^{11}$	$< 3.6 \times 10^{12}$
μ^-	muon	105.7	1.23×10^{12}	1.23×10^{13}
$\pi^\pm$	pion	139.6	1.62×10^{12}	1.62×10^{13}
p, n	proton, neutron	938.6, 339.6	1.09×10^{13}	1.09×10^{14}
τ^-	tau	1784	2.07×10^{13}	2.07×10^{14}
$W^\pm$	W particle	80220	9.31×10^{14}	9.31×10^{15}
Z^0	Z particle	91187	1.06×10^{15}	1.06×10^{16}

the photon energy, hf, and the variables are changed from momentum to frequency, f, via

$$p = \frac{hf}{c} \tag{5.9}$$

As the Universe expands and cools, the kinetic energy possessed by particles of matter will also fall. Eventually, the particles will cease to travel at relativistic velocities, and the distribution becomes

$$N = n_{spin} \frac{(2\pi mkT)^{3/2}}{(ch)^3} e^{-E/kT} \tag{5.10}$$

where N is the number density of non-relativistic particles with mass, m, and spin states, n_{spin}, at a temperature T. This is known as the Maxwell–Boltzmann distribution. In general, a particle is relativistic when its kinetic energy is 10 times larger than its rest mass energy (see Table 5.1).

The rest mass energy of a particle is very important in a cosmological context because, when the energy density of radiation in the Universe is greater than twice this figure, the particle and its antiparticle can be spontaneously created. Because the antimatter particle must always be created, the energy density must be twice the rest mass energy of the particle to be created. The excess energy – over and above what is needed to create the particle – is received by the particle as kinetic energy. Thus, at energies greater than 20 times the particles' rest mass energy, the particles are created with relativistic velocities. As the temperature – and hence the energy density – of the Universe drops, so each individual particle species is created with less and less kinetic energy until, finally, production stops, because the Universe has dropped below the threshold temperature. When this happens, two results are possible. In the vast majority of cases, the particle species ceases to exist in any appreciable number, because its particles all annihilate with the antiparticles. In the cases of particles such as neutrons, protons and electrons, some process in the early Universe (which we shall discuss later) allowed the creation of particles without antiparticles. Thus, when the Universe passed their threshold temperatures, not all of the particles were destroyed, because of a deficiency in the number of antiparticles. These remaining particles – which have gone on to form all of the stars and galaxies around us – were non-relativistic, and all followed Maxwell–Boltzmann distributions.

As we begin at the earliest time in history about which physics can say anything, the evolution of the Universe is driven by its expansion and its effects upon the energy density and temperature. Hence, in the following sections we will frequently refer to the temperature and/or the energy of the Universe, at the most important events in its history.

5.2   THE FIRST 10^{-43} SECONDS

Modern-day physics can investigate the Universe back to a fraction of a second after the Big Bang, but no earlier. The earliest age at which conventional physics can say anything at all about the Universe is approximately 10^{-43} s. This is a special interval in time known as the Planck time, which arises from the fledgling quantum theories of gravity. But why search for a quantum theory of gravity when (as we saw in Chapter 1) general relativity can describe it?

Even Albert Einstein refused to believe that his general theory of relativity was complete. Late in his life, he questioned the validity of his theory for certain astrophysical circumstances. He pointed out that general relativity was intended only as a valid approximation for weak gravitational fields. In formulating his theory, he assumed that he could separate the concepts of matter and gravitational field. In a region of space where the matter density was incredibly high, Einstein felt that his theory would break down.

Nowhere can there be a more severe test of this than at the beginning of the Universe, when everything was squeezed together, producing the most incredibly high densities. In the same way that Newtonian physics must defer to relativistic physics when objects travel faster than 10% of the speed of light, so there is a size limit below which a quantum theory must be used to explain the behaviour of matter. For example, when discussing a planet such as the Earth in orbit around a star such as the Sun, the Newtonian approximation or general relativity are capable of being used. If the scale is changed dramatically, and an electron in orbit around a hydrogen nucleus is considered, quantum corrections will become highly significant, and the previous two theories are inadequate for the task.

Beginning with equation (3.3) – the Heisenberg uncertainty principle – it can be shown by dimensional analysis that the Planck time, t_p, is represented by

$$t_P \cong \sqrt{\frac{hG}{2\pi^5}} \approx 10^{-43} \qquad (5.11)$$

This length of time is of fundamental importance in our consideration of the early Universe. It is so important that it can be used to form a series of fundamental units, including the Planck mass

$$m_P \cong \frac{h}{2\pi c^2 t_P} \cong 2.5 \times 10^{-5} \text{ g} \qquad (5.12)$$

and the Planck length

$$l_P \cong ct_P \cong 1.7 \times 10^{-33} \text{ cm} \qquad (5.13)$$

The reason why this system of units has become used, and why the Planck time is so important, will becomes obvious with the following analysis.

In Chapter 2 we introduced a quantity known as the Compton wavelength (equation (2.7)), which represents the wavelength a photon would possess if it were to carry the rest energy of a particle. We have also used the uncertainty principle (equation (3.3)), to determine the time for which a gauge boson could exist (equation (3.4)). This equation can be used in conjunction with any particle to determine the Compton time, t_c. This time is very important, because it represents the length of time for which the mass, m, can violate the conservation of energy law:

$$t_c = \frac{h}{2\pi mc^2} \qquad (5.14)$$

This kind of behaviour is called a quantum fluctuation. It is usual for it to take the form of pair production. A pair of virtual particles (one a particle of matter, the other its antimatter counterpart) are spontaneously created, but annihilate themselves, returning the borrowed energy before they can be detected. This all takes place in a time of t_c. The Compton time can be easily converted into a Compton length, which is virtually indistinguishable from the Compton wavelength. The Compton length is the typical distance travelled by the virtual particles before they annihilate each other:

$$l_c = ct_c \qquad (5.15)$$

The Compton length can therefore be envisaged as the limiting distance for any particle in the quantum regime. In classical physics, a similar type of limiting distance can also be derived. It is known as the Schwarzchild radius, and is the classical definition of the radius of a black hole which possesses a mass, m. It is represented by

$$l_s = \frac{2Gm}{c^2} \qquad (5.16)$$

Nothing can escape from within a radius of l_s around an object of mass m. Thus, l_s is the boundary between the classical regime of gravitation and quantum gravity. For convenience with comparisons, let us define a Schwarzschild time, t_s, which is simply the time that a photon of light would take to travel the Schwarzschild radius:

$$t_s = \frac{l_s}{c} \tag{5.17}$$

These times and distances can be used to determine if we are in the classical or quantum realm. If the Compton length and Schwarzschild radius of a particle with the Planck mass are calculated, it will be noticed that the calculated lengths are both equal to the Planck length. Converting these quantities to time, both Schwarzschild and Compton time are equal to the Planck time. In other words, a particle of the Planck mass is a minuscule black hole which briefly exists for 10^{-43} s – the Planck time. If the mass of the particle is increased, then it enters the realm of macroscopic entities. In this case, $t_c < t_s$, $l_c < l_s$ and quantum effects can be neglected. In particular, the self-gravity of the body – its gravitational effect upon itself – can be described using relativity or, if approximations can be applied, Newtonian theory. For particles with a mass less than that of the Planck mass, however, $t_c > t_s$, $l_c > l_s$ and quantum effects are of crucial importance. It is no longer adequate to describe, with general relativity, the self-gravity of a body possessing $l_c > l_s$. As we have seen, this is the realm of the black hole – a body whose self-gravity is obviously of the highest importance, because it has crushed itself out of existence! A theory of quantum gravity is necessary to fully describe this behaviour, and although many physicists are currently working towards this, no-one has convincingly presented anything which even resembles a finished theory.

Thus, on time-scales of the Planck time, black holes of the Planck mass can spontaneously come into existence! Via the process of Hawking radiation, the black hole can then evaporate back into energy. The characteristic time-scale for this to occur happens to be approximately equal to the Planck time. Thus, the Universe at 10^{-43} s in age was a seething mass of energy in which black holes were continuously forming and evaporating. The concept of distance (a measure of the amount of spacetime continuum between two events) can have very little meaning in the Universe at this stage. The continuum was squeezed into such a tiny volume of space that regions of space which would end up as being totally disparate were in contact. This is best imagined as being like a flat sheet of paper which began as a screwed-up ball. Many different parts of the paper would be in contact with one another, but as the paper is flattened so they will lose contact.

The process of pair production is thought to have been incredibly important at these early times because, by the action of a strong enough electromagnetic field or a sufficiently non-uniform gravitational field, it is possible to separate the two particles before they recombine. The gravitational field which permeated the Universe, at this stage in its history, was exceptionally non-uniform. Hence, many of the particles forged in pair production escaped their initial annihilation. The density of these particles was so high that they soon met and annihilated with other antiparticles. This returned their bound-up energy back into photons. The Universe, however, had begun to be seeded with matter.

Thus, the age of 10^{-43} s was the first important watershed in the Universe's history. At this time, gravity was just becoming an individual force, separate and distinct from the other three which were still joined together in a grand unified force.

The Planck era (as the time before is known) was the realm of the superforce, and cannot be explored with our current theories. Between the Planck era and the next watershed, which occurred at 10^{-35} s, the quantum gravitational effects dwindled to become negligible. From that age onwards, general relativity and the Newtonian approximation become a cosmologist's greatest allies in understanding the evolution of the spacetime continuum.

5.3 THE PHASE TRANSITION ERA

The separation of gravity from the superforce marks the first known phase transition in the Universe. As indicated in the previous section, physics can explain very little about this event because of the lack of a suitable theory. The first phase transition which can be explored theoretically is that which occurred when the age of the Universe was 10^{-35} s. This is when the grand unified force split up to form the strong nuclear force and the electroweak force.

The way in which particles behaved and interacted with one another during this epoch of the Universe's existence can be described by a grand unified theory (GUT). There are many approaches and, consequently, many different versions of these at present. None have been proved, and so the behaviour of matter during this phase of the Universe is still a matter for conjecture.

During the GUT era the strong nuclear force, the weak nuclear force and the electromagnetic force all acted as one grand unified force. There was therefore no distinction between hadrons and leptons. All particles behaved in the same way, and interacted through the exchange of the same virtual particles. In cosmological parlance, these particles were said to possess symmetry.

Symmetry is a word that is used to define a set of transformations which can be applied to a system without altering any of its physical properties. For instance, if we briefly switch to a much larger scale, molecules exhibit reflectional and rotational symmetry (see Figure 5.2). The symmetries studied in elementary particles are often more esoteric, and are not so easy to relate to physical properties. Perhaps the best way to envisage it is that particles which obey a symmetry are indistinguishable from one another. A branch of mathematics known as group theory is used to study these properties.

In general, the Universe today is a place of broken symmetries. There are many different particles, all of which act differently from one another. In the high-energy phase of the Universe – such as before the age of 10^{-35} s – the symmetries were largely unbroken. The particles were said to obey the grand unified theory symmetry. According to this theory, it is here, at this stage in the Universe's history, that the propensity of matter over antimatter was introduced into the Universe. The grand unified theories make leptons and hadrons equivalent; therefore our modern notion of baryon conservation in reactions (which results in particle–antiparticle pair production) no longer holds true. In particular, certain exotic particles could be formed which would later decay into a proton without its matching antiproton. The proton would then be broken up immediately into quarks. Conditions which allow

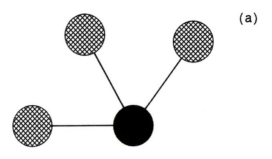

This molecule has the same physical
properties as its mirror image:-

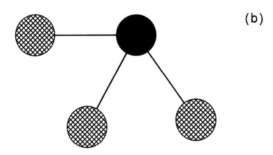

Figure 5.2. Molecules can display reflectional and rotational symmetry. A molecule behaves in exactly the same way, regardless of its rotational state of symmetry. At a much smaller level, different species of sub-atomic particles are believed to behave in the same way at high energies. This concept is known as symmetry.

the occasional creation of a baryon without an antibaryon counterpart are therefore possible, under this theory, in a (calculated) 1 in every 10^8–10^{13} cases. This process is sometimes referred to as baryosynthesis. It is still hotly debated about where, in time, baryosynthesis took place. Some cosmologists believe that it was here in the separation of the grand unified force that the asymmetry was introduced, whilst others believe it occurred later, at about 10^{-10} s.

When the grand unified symmetry spontaneously broke because the temperature had reached 10^{28} K – corresponding to an energy of 10^{15} GeV – hadrons and leptons became separate entities. The strong nuclear force began to behave in a manner which was distinct from the other two, which were still joined by the electroweak force. An excellent example of spontaneous symmetry breaking occurs in cooling ferromagnets. At temperatures above the Curie temperature, the internal particle arrangement is such that no magnetic field is produced. In this case, the ferromagnet could be oriented in any way with no change in any of its external properties. It can therefore be said to have a high level of symmetry. As the temperature drops below

Magnet above the Curie temperature:

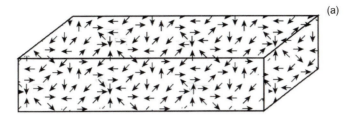

(a)

Dipoles are totally random, no overall magnetism

Magnet below the Curie temperature:

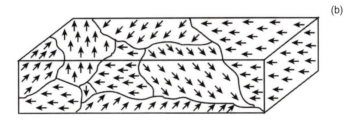

(b)

Regions of dominant magnetism have now spontaneously formed

Figure 5.3. Spontaneous creation of magnetism. (a) Above a certain temperature a ferromagnet is not globally magnetised, because the magnetic dipoles are randomly oriented. (b) When the temperature drops below the Curie temperature, the magnetic dipoles spontaneously align, causing domains of overwhelming magnetism.

the Curie temperature, the internal particles spontaneously arrange themselves so that magnetisation occurs. Suddenly re-orienting the ferromagnet will change the way the outside world perceives it, and thus its earlier symmetry is been broken (see Figure 5.3).

In the process of spontaneous symmetry breaking in the ferromagnet, the final direction in which the individual dipole atoms find themselves is totally arbitrary. One region of the magnet will be magnetised in one direction, whilst other regions will be magnetised in a different direction. Each region of dominant magnetisation is called a domain, and between these are boundaries of magnetic discontinuities.

At the time of the grand unified force's break up, a similar event is postulated to have occurred in our Universe. These discontinuities could present themselves in one of four different styles: a single point defect is known as a magnetic monopole; a one-dimensional line discontinuity is termed a cosmic string; a two-dimensional discontinuity goes by the name of a domain wall; and a three-dimensional discontinuity is called a texture. So far, none of these have been observed.

Our discussion of hadrons at this stage in the Universe's evolution almost exclusively refers to individual quarks. In our Universe today, quarks are bound up inside hadrons, but in the high-energy Universe of this cosmic era the quark–hadron symmetry had yet to be broken, and so quarks existed as single, individual particles. Any protons which were formed were instantly broken up into their constituent quarks.

Another symmetry began to be broken when the temperature of the Universe had dropped to 10 K (100 GeV). In the process of this symmetry spontaneously breaking, the electromagnetic force separated from the weak nuclear force. This began to occur when the Universe had reached an age of approximately 10^{-12} s. By the time the temperature had fallen to 10^{13} K (1 GeV) this phase transition was complete. During this process, leptons acquired mass. A final phase transition took place at a temperature of approximately 2×10^{12} K (200 MeV). It marks the beginning of the hadron era, about which cosmologists become rather more confident about the physical processes which were taking place. Everything until this point in history must be regarded with a healthy level of scepticism. Until the technology of particle accelerators increases sufficiently to allow these energies to be more closely studied, details and perhaps even fundamental ideas might be liable to change.

5.4   THE HADRON AND LEPTON ERAS

At this point in the history of the Universe (200 MeV) the quark–hadron symmetry finally broke. The broken symmetry affected the strong nuclear force, and acted in such a way that it confined the individual quarks into mesons and baryons. Thus, as well as the leptons and photons which dominated the Universe, protons, neutrons, their antiparticles and pi-mesons (pions) were created, replacing the free quarks and antiquarks.

The hadron era of the Universe did not last for very long. The strengthening of the inter-quark component of the strong nuclear force rendered it impossible for quarks to exist on their own. When a proton annihilated with its antiproton, the vacuum energy was too low for it to be replaced, since the threshold energy for the spontaneous creation of a proton–antiproton pair is about 2 GeV, which had long since been passed. Thus the baryon content of the Universe inexorably dwindled, until all that was left was the tiny residue of matter created during the grand unified epoch. For a short time, pions dominated, until their threshold energy was also passed. Three types of pion are possible; they can be either positively or negatively charged, or they can be neutral, depending upon their precise quark content. At this point, the charged pions annihilated each other and were not replaced, whilst the neutral pions decayed into photons and were not replaced. Pion decay took place when the vacuum energy dropped below 130 MeV, signalling the end of the hadron era. The photon density of the Universe was increased enormously because of all the matter–antimatter collisions. The residual hadrons remained in thermal equilibrium with the leptons and photons. The temperature in the Universe at this point was approximately 10^{12} K.

The lepton era lasted for the period of time between the vacuum energies of 130–0.5 MeV. It began when the age of the Universe was 10^{-5} s. The threshold energy of the electron is 1 MeV, whilst its heavier cousins – the muon and the tau – have threshold energies which are higher. The tau may actually be so massive that it had already passed its threshold temperature during the hadron era. Whenever the exact time, the first to annihilate were the taus, followed by the muons at 10^{12} K.

The annihilation of the muons had important repercussions for the neutrinos. Prior to this, the density of particles had been so great that the neutrinos could travel only relatively short distances before interacting with other particles. Neutrinos interact with other particles via the weak nuclear force, which means that they have to pass within a distance of 10^{-17} m before they can exchange gauge bosons and exert any influence upon one another. In the modern-day Universe, approaching this close to another particle is not easy because matter is so dissipated. Even in so-called 'solid' objects there is a large amount of space on these kinds of scales. The consequence of this is that neutrinos can pass through solid objects such as this book, the reader and even the Earth itself, as if they were non-existent. In the early Universe, however, particles were squeezed so closely together that neutrinos were constantly passing other particles closer than 10^{-17} m. Hence, they interacted regularly and maintained thermal equilibrium. With so many particles of matter being lost in antimatter annihilations, the density of matter dropped precipitously. The annihilation of the muons proved decisive for the neutrinos. Suddenly the matter density was so low (by neutrino standards) that they could travel long distances without interacting with other particles – a condition known as decoupling.

An important consequence of decoupling was that the neutrinos began to depart from the thermal equilibrium of the rest of the Universe. They began to lose energy independently, because they no longer collided with other particles. Neutrino decoupling took place at 3×10^{10} K. Another particle which can interact via the weak nuclear force is the neutron. Since the creation of hadrons in the previous cosmic era, neutrons have been constantly interacting with the surrounding particles via the reactions

$$e^- + p \Leftrightarrow v_e + n$$
$$\bar{v}_e + p \Leftrightarrow e^+ + n \tag{5.18}$$

As can be seen, reactions have sometimes produced neutrons, whilst at other times they have changed the neutrons into protons. Each time, however, the reaction has involved a neutrino – and that is the signature of the weak nuclear reaction.

When the neutrinos decoupled, the reaction rate of the weak nuclear force effectively plummeted. When compared to the reaction rates of the other fundamental forces, it is almost possible to discount the weak nuclear force from further consideration, so it can be said that at this point in the Universe's history the number of neutrons and neutrinos became fixed. With sufficiently sensitive neutrino detectors, it should be possible to detect this neutrino background. The number of neutrons had very important consequences for the next stage in the Universe's evolution

At the end of the lepton era the content of the Universe was finally in its present-day proportions, with photons vastly outnumbering the particles of matter. Interactions were still so common between the nucleons, electrons and photons that all remained in thermal equilibrium. The nucleons and electrons no longer equally participated in the thermal evolution of the Universe, however, because they were vastly outnumbered by the photons. This is why this event signifies the beginning of the radiative era of the Universe. This should not to be confused with the term 'radiation-dominated Universe'; the Universe ceases to be radiation-dominated very shortly after this event. During the radiative era the photons vastly outnumbered the matter particles, and alone drove the thermal evolution of the Universe. In the radiation-dominated Universe the relativistic particles (the radiation) govern the expansion of space.

5.5   THE NUCLEOSYNTHESIS OF HELIUM

Chapter 4 discussed the subject of matter or radiation dominance in the Universe. In order to define these cosmic epochs, the energy density of radiation (equation (5.5)), must be equal to the energy density of matter, ε_m. This is determined to first approximation by dividing the rest mass energy (equation (1.2)) by the volume, V:

$$\varepsilon_m = \frac{Mc^2}{V} \tag{5.19}$$

This term neglects any kinetic energy that the particles of matter may possess. This is not disastrous, because at the epoch in which matter becomes dominant, any kinetic energy the particles possess will be small compared to their rest mass energies. As stated in Chapter 4, the dividing line occurred at roughly 100 s after the Big Bang. This is so close to the beginning of the epoch of nucleosynthesis that the energy density of the Universe ceased to be radiation-dominated and became matter-dominated when nucleosynthesis began. At this point there was a change in the way in which the Universe expanded.

Nucleosynthesis began when the temperature of the Universe dropped to approximately 10^9 K. This was the situation that occurred at a time of approximately 1.5 minutes after the Big Bang – around the time that the Universe became matter-dominated. The number of neutrons and protons was not equal, because neutrons are more massive than protons. The Maxwell–Boltzmann distribution (equation (5.10)) can be used to determine the ratio of neutrons and protons at that time.

$$\frac{N_n}{N_P} = \left(\frac{m_n}{m_P}\right)^{3/2} \exp\left(-\frac{m_n - m_P}{kT}\right) \tag{5.20}$$

At this stage in the history of the Universe it was about 0.1–0.2: protons outnumbered neutrons by 5–10:1. Neutrons in isolation are not stable particles, and will radioactively decay into a proton via the emission of an electron (beta decay). The mean lifetime of a free neutron is about 14.8 minutes but, if it can become bound into an atomic nucleus, the neutron will no longer decay. It will remain stable for as long as it is contained within the nucleus.

Chapter 1 introduced the proton–proton chain by which helium is synthesized in the hearts of stars. During the helium nucleosynthesis of the early Universe the initial collision between two photons was largely replaced by the collision of a proton and a neutron, which forms a deuterium atom and a photon of energy.

$$\text{}^1_1\text{H} + \text{n} \rightarrow \text{}^2_1\text{H} + \gamma \tag{5.21}$$

Nuclear fusion introduces a quantity known as the binding energy – the quantity of energy which is carried away by the photon in an exothermic fusion reaction. For example, in the nuclear fusion between a neutron and a proton, the resulting deuterium nucleus is less massive than the combined mass of the proton and the neutron. This difference is the mass deficit, Δm, and when the energy equivalent is calculated using the Einstein equation (1.2), this is the quantity known as the binding energy.

A deuterium nucleus has a binding energy of 2.22 MeV; as long as there are photons containing that amount of energy present in the Universe, the reaction could be reversed. When a deuterium nucleus collides with a 2.22 MeV (or higher) photon, it will be broken up into a neutron and a proton in a process called photodisintegration. Even when the energy density of the Universe drops considerably below 2.22 MeV, there is still a considerable number of these high-energy photons which can photodisintegrate the newly formed deuterium nuclei. So, despite the formation of deuterium being favoured when the energy density dropped to its nuclear binding energy, the photodisintegration stopped large numbers of the nuclei persisting until the energy had fallen still further to 0.07 MeV.

This delay in the accumulation of deuterium nuclei is referred to as the 'deuterium bottleneck'. Once past this hold-up, fusion began in earnest, with reactions building upon the deuterium 'building bricks' to forge quantities of another hydrogen isotope, tritium, which contains two neutrons and possesses a binding energy of 8.48 MeV. Also favoured was the formation of a light isotope of helium – helium-3 – containing two protons but only one neutron, with a binding energy of 7.72 MeV. Helium nuclei are then formed from combination of neutrons, protons, tritium and helium-3. The resulting helium nuclei – which have two protons and two neutrons – are extremely stable, because they have binding energies of 28.3 MeV. In this way, during the epoch of nucleosynthesis every neutron was bound up into a nucleus before the majority of them could decay into protons and electrons.

The rate at which a fusion reaction, r, takes place depends upon the density of reacting particles and the Gamow penetration factor – a function of the energy bath in which the particles are immersed, E, and the charges of the two reacting nuclei, Z_1 and Z_2:

$$r \propto \exp\left(-\frac{2Z_1Z_2}{\sqrt{E}}\right) \tag{5.22}$$

As E decreases, r also decreases. Consequently, as the Universe expanded and the energy density dropped, nuclear reaction became less and less likely. Even more serious was that, as Z_1 and Z_2 increase, r decreases. So, as the nuclei became more and more charged (accumulated protons) the reaction rate dropped rapidly. These two effects combined to virtually halt nuclear fusion in the early Universe after the neutrons had been confined in helium nuclei. Only minute quantities of the element lithium were formed from the helium. Fusion stopped when the Universe reached approximately five minutes in age and the temperature had dropped to around 6×10^8 K (52 keV).

It was therefore during this epoch that the primordial cosmic abundance of elements was forged. 25% of the mass of the Universe was helium, whilst an enormous 75% remained as hydrogen. There were also very small trace amounts of deuterium, helium-3 and lithium.

5.6   THE DECOUPLING OF MATTER AND ENERGY

After the frantic evolution of the previous five minutes, the Universe entered a state of relatively calm evolution. It continued to expand, and the energy density and the temperature dropped accordingly. The temperature of the Universe was then governed by the photons of radiation, which vastly outnumbered the particles of matter. At the end of the nuclear fusion epoch the ratio of matter to radiation was set in its current-day ratio of one particle of matter to every 2,000 million photons.

As well as containing this vast amount of photons, the Universe at this stage in its evolution was filled with a high-energy plasma, consisting of the nuclei forged during the epoch of nucleosynthesis, and a sea of free electrons. The hydrogen nuclei themselves were nothing more than single protons.

The number density of particles was still so high that they were all in a constant state of collision. When photons of electromagnetic radiation cannot travel very far through a medium without colliding with a particle of matter, the medium is said to be optically thick. In the Universe today, space is optically thin because light can travel across vast distances of space virtually unhindered. This is why we can see distant galaxies. In the Universe before the decoupling of matter and energy, conditions were optically thick.

In this high-energy environment, energy was being shared between the particles via the bremsstrahlung process (described in Chapter 2), creating new photons to carry away excess energy. Also, photons were interacting with matter via Compton and inverse Compton scattering. This high level of interaction meant that both matter and radiation were maintained at the same temperature – a process known as thermalisation.

When the Universe was one year old, the expansion of space had lowered the density of matter sufficiently so that bremsstrahlung ceased to be a dominant process. Thermalisation stopped, and the temperature of matter began to evolve differently from the temperature of radiation. The expansion of the Universe stretched the radiation to successively longer and longer wavelengths. But this did not affect the shape of the photon energy distribution; it still followed a black-body curve. As the wavelengths were lengthened, so the temperature of the radiation fell.

The particles of matter began to evolve under the influence of gravity, which caused them to begin clumping together. In this way, certain regions of the Universe became slightly denser than the average, whilst other regions were subjected to a drop in density. Radiation continued to interact with matter via the Compton effect, and so these differences in density began to impose an influence on the radiation, which was no longer constant throughout the whole Universe.

For the next 300,000 years the Universe remained optically thick to the radiation. Matter continued to slowly clump together, and the radiation cooled according to the expansion of the Universe (with a slight influence from the density of the region in which the photons existed). The turning point between an optically thick and an optically thin Universe was the next major development in cosmic history, and it signalled the end of the plasma epoch (see Figure 5.4).

The plasma which permeated the Universe continued to be ionised. As stated in Chapter 2, the process of ionisation is a bound–free transition which liberates an electron from its electrostatic attraction to an atomic nucleus. In the case of a hydrogen atom, an energy of 13.6 eV is necessary to ensure the removal of the electron. As electrons continued to move around, the few collisions they did suffer inexorably drained them of kinetic energy, until they could be easily captured by any atomic nucleus if they happened to stray too close. The atomic nuclei – being much more massive than the electrons – moved very slowly in comparison. By this stage, photon energies had dropped considerably below 13.6 eV, and so when the electrons were captured they could no longer escape. Instead of high-energy electrons moving erratically around the Universe, they were confined to the near vicinity of atomic nuclei.

By removing the interference of the electrons, photons suddenly became able to travel large distances without interacting (see Figure 5.5). The Universe became optically thin at this stage, when the temperature of the radiation was approximately 3,000 K (energy density 0.26 eV). This is the point at which matter is said to have decoupled from energy. The photons of radiation, which suddenly began to travel large distances through the Universe, are the same photons that we observe today as the microwave background radiation. The temperature of that radiation has been reduced by a factor of 1,000, due to the expansion of the Universe. Slight differences in matter density from place to place in the Universe resulted in slight differences of temperature in this radiation. It was these temperature variations that were recently detected by COBE.

5.7   MODELLING THE EXPANDING UNIVERSE

The previous sections in this chapter have shown that the driving force behind the Universe's evolution is its expansion. In equation (5.4) we introduced a function $R(t)$, which characterised the linear size of the Universe. Thus, as time passes $R(t)$ grows in size for an expanding Universe. This is a function known as the scale factor, and we shall often refer to it from now on. Equation (5.4) showed how the volume of the Universe depends upon $R(t)^3$.

Chronology of the Big Bang

Era	Time (s)	Temp. (K)	Notes
Planck Era	$t = 0$		The Big Bang occurs.
			All four fundamental forces unified; physics of the Universe indeterminate.
	$t = 10^{-43}$	10^{32}	Gravity separates from other three forces.
GUT Era			
	$t = 10^{-35}$	10^{28}	Strong nuclear force separates from electroweak force. GUT symmetry breaks (hadrons and leptons no longer equivalent). Baryosynthesis takes place. Inflationary epoch begins. (Universe increases its size by ~ 10^{50} times).
Electroweak Era			
	$t = 10^{-11}$	10^{15}	Electroweak symmetry breaks. Leptons acquire mass.
Hadron Era	$t = 10^{-10}$	10^{12}	Quark–hadron symmetry breaks. Protons and neutrons are created.
Lepton Era	$t = 10^{-5}$	10^{10}	Neutrinos decouple.
	$t = 10^{2}$	10^{9}	Nucleosynthesis begins.
Radiative Era	$t = 10^{2.5}$	10^{8}	Nucleosynthesis ends.
	$t = 10^{7.5}$ (t = 1 year)		Thermalisation between energy and matter ceases.
	$t = 10^{13}$ (t = 300,000 years)	10^{3}	Decoupling of matter and energy. Universe becomes transparent to photons; these photons observed today as cosmic microwave background radiation.

Radiation-dominated Universe

Matter-dominated Universe

Figure 5.4. Key events in the evolution of the Universe just after the Big Bang.

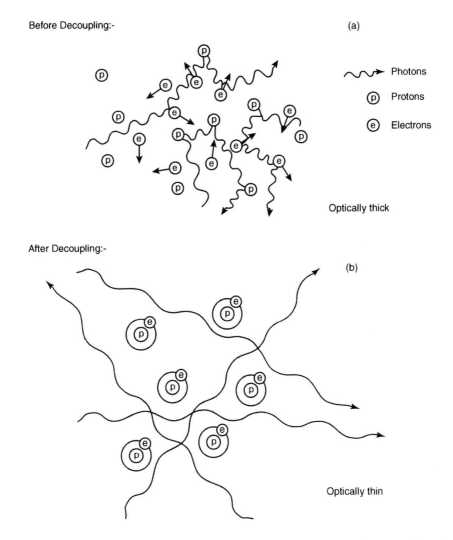

Figure 5.5. Optical thickness before and after decoupling. (a) Before the decoupling of matter and energy, photons of radiation could not travel far before colliding with particles of matter. (b) After the decoupling, the electrons were trapped by the atomic nuclei, and the photons could travel for large distances without interacting.

We have seen how the strong and weak nuclear forces act only over the distances of atomic nuclei and atoms. Chapter 3 also told how electromagnetic forces average out to be zero over large volumes of space. Therefore, the expansion of the Universe since a time of 1 µs has been governed exclusively by gravity. In Chapter 4 we remarked that Newton wondered why the Universe did not collapse through the gravitational pull of its constituents, because he thought that the Universe was static. The potential to be moved by gravity can be quantified, and is known, appropriately,

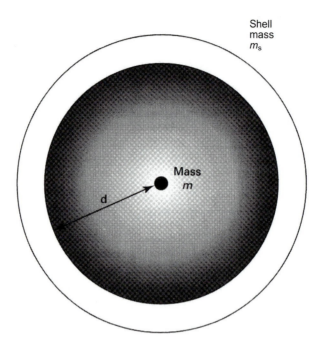

Figure 5.6. The Friedmann equation can be derived by considering a sphere of the expanding Universe and exploring what would happen to the matter at the extreme edges of the sphere. (Adapted from Silk, J., *A Short History of the Universe*, W.H. Freeman, 1994.)

as gravitational potential energy. Hubble discovered that the Universe was expanding, and solved Newton's problem. Anything which moves is said to have kinetic energy. Hence, in our Universe the kinetic energy of the separating galaxies is greater than the force of gravity between them. This may not always remain the case, however.

In order to model the expansion of the Universe, this inequality in energies must be formulated by mathematics. In order to do this we imagine an arbitrary spherical volume of the Universe, and overlay a thin spherical shell around it. The sphere contains a mass, m, whilst the shell contains a mass, m_s (see Figure 5.6). The sphere has a radius, d, whilst the thickness of the shell is negligible. In accordance with the expansion of the Universe, the matter in the shell is travelling away from the centre of the sphere with a velocity, v, given by the Hubble expansion law, equation (1.4). With these as our basic parameters, the kinetic energy can be calculated.

$$\text{K.E.} = \frac{1}{2} m_s v^2 \tag{5.23}$$

The gravitational potential energy can also be calculated.

$$\text{P.E.} = -\frac{Gmm_s}{d} \tag{5.24}$$

Together, these can be equated to find the total energy of the shell by application of the principle of energy conservation, which states that

$$\text{K.E.} + \text{P.E.} = \text{total energy} \tag{5.25}$$

Before using equation (5.25) to find the total energy of the shell, we shall substitute for the velocity in equation (5.23) by using equation (1.4), we shall also substitute for the mass of the sphere in equation (5.24) by using the equation

$$m = V\rho = \frac{4}{3}\pi d^3 \rho \tag{5.26}$$

Thus equation (5.25) becomes

$$m_s \left(\frac{H(t)^2 d^2}{2} - \frac{4G\pi\rho d^2}{3} \right) = \text{total energy} \tag{5.27}$$

This equation includes the density of the Universe, ρ, which is proportional to the inverse cube of the scale factor, $R(t)^{-3}$, and hence drops by a factor of eight if the Universe doubles its linear dimensions (see Figure 5.7). If the total energy, E, is replaced with an energy constant, k, equal to

$$k = \frac{2ER(t)^2}{m_s d^2} \tag{5.28}$$

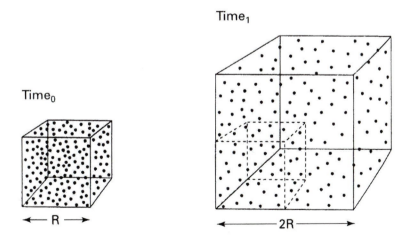

Figure 5.7. As the Universe expands, the volume increases in proportion to the cube of the scale factor. (Adapted from Silk, J., *A Short History of the Universe*, W.H. Freeman, 1994.)

then equation (5.27) becomes

$$H(t)^2 - \frac{8\pi G\rho}{3} = \frac{k}{R(t)^2} \tag{5.29}$$

This is known as the Friedmann equation, and describes the expansion of the Universe in terms of measurable quantities. These are the Hubble constant, $H(t) = H_0$, the present density of the Universe, $\rho = \rho_0$, and the present scale factor, $R(t) = 1$, which we can set to unity because it is an arbitrary measure of the size of the Universe. Thus, if we can determine the Hubble constant and the density of the Universe we could solve this equation and find k. Unfortunately, these parameters cannot be easily determined (as will be seen in later chapters). The actual magnitude of k is unimportant. The big question is: is it positive or negative?

The kinetic energy term in the Friedmann equation (5.29) has reduced to the square of the Hubble constant, which is traditionally thought to decrease with time because of the action of gravity. The potential energy is governed by the density of the Universe. The density of the Universe also decreases as the Universe expands but, because the sign of the potential energy term is negative, this results in an increase in the gravitational potential energy.

The sign of k depends upon whether the magnitude of the gravitational potential energy is greater than the kinetic energy. If it is, the value of k is negative. If not, then k becomes positive. At the point where the kinetic energy is just equal to the potential energy, k becomes 0.

When the Friedmann equation is derived through use of Einstein's general relativity, the energy constant is interpreted differently. Instead of an energy, it is thought to represent the overall curvature of the spacetime continuum. Open Universes correspond to positive values of k, and are analogous to the spacetime continuum being shaped like a saddle. Closed Universes correspond to negative values of k, and are analogous to spheres. When $k = 0$ the Universe is said to be flat, (see Figure 5.8). The consequences of these curvatures and their dependence upon the density of the Universe and the Hubble constant will be discussed fully in the final chapter, when we consider the age and the fate of our Universe.

The Big Bang theory is without doubt the best current interpretation of the cosmological observations available, but it is by no means proven. Within its structure there exist a number of problems which may or may not prove to be solveable. As the final section in this chapter shows, the theory has already undergone one major revision, and it seems likely that others will follow as better data become available and better theoretical ideas are formulated.

5.8   THE FLATNESS AND HORIZON PROBLEMS

The flatness problem arises when present-day estimates of the k term in the Friedmann equation are attempted. This is enveloped in the search to discover the density of the Universe – as discussed in Chapter 11 – but for the moment it is

In a positively curved
Universe the angles
inside a triangle add up
to more than 180°

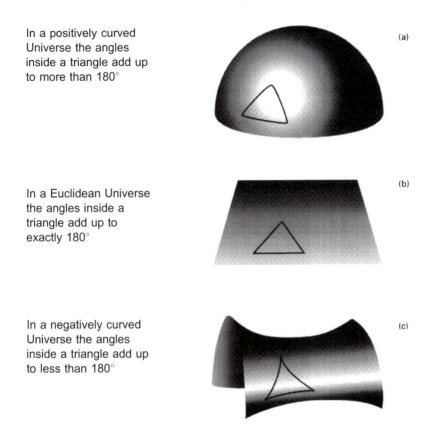

In a Euclidean Universe
the angles inside a
triangle add up to
exactly 180°

In a negatively curved
Universe the angles
inside a triangle add up
to less than 180°

Figure 5.8. The curvature of the spacetime continuum is dependent upon the amount of matter within it. (a) The closed Universe contains enough matter to halt the expansion and collapse the Universe. (b) The 'flat' (Euclidean) Universe contains enough matter to halt the expansion, but only after an infinite length of time. (c) The open Universe does not contain enough matter to halt its expansion, and so will expand forever. (Adapted from Kaufmann, W.J., *Universe*, W.H. Freeman, 1987.)

enough to simply state that all analyses determine a value for k which, within the bounds of experimental accuracy, is equal to 0 and implies a flat geometry for the spacetime continuum. Simply stated, the flatness problem is: why should we live in a flat Universe? An infinite number of open (k > 0) and closed (k < 0) Universes are possible; so why, out of all the values which k could take, is ours so close to the one special case – the dividing line between an open and a closed Universe? It seems to be too great a coincidence; something that requires further investigation. Coincidences usually lead scientists to a reason why things are as they are.

Chapter 4 introduced the concept that any event has an horizon whose radius is determined by the time since the event multiplied by the speed of light. At the decoupling of matter and energy the Universe is thought to have been 300,000 years old, and so its horizon was about 300,000 light years in radius. It will be shown in Chapter 11 that the Friedmann equation can be used to demonstrate that the dependence of scale factor with time for a flat Universe is

$$R(t) \propto t^{2/3} \tag{5.30}$$

If we assume that the current age of Universe is 1.5×10^{10} years, then the current horizon is 1.5×10^{10} light years. Using equation (5.30) to calculate the present scale factor and the scale factor for the Universe at the time of decoupling, it can be easily calculated that the size of the Universe at 300,000 years is about 1,350 times smaller, which in terms of an actual distance is about 1×10^7 light years. So, at the time of decoupling, the presently observable Universe was at least 30 times bigger than the horizon distance at that age.

This leads to a problem with the microwave background radiation which has become known as the horizon problem. As described in Chapter 2, the CMBR possesses the same properties, no matter in which direction it is detected. Although there is the dipolar anisotropy produced by the motion of the Earth though space, and the fluctuations produced by tiny variations in the local density of matter, in principle the radiation is entirely smooth. It possesses the same black-body temperature regardless of it direction. Conventionally, this behaviour is possible only if the radiation throughout the entire volume of the presently observable Universe were to be in thermal equilibrium. The only way in which thermal equilibrium could be maintained is if energy could be shared between all particles of matter. This was clearly impossible at the time of decoupling, because even photons (which travel at the speed of light) could not have travelled the required distances to share the energy.

The horizon problem and the flatness problem led to the proposal by Alan Guth, in 1980, that the Universe entered a period of exponential expansion during its early history. Although his ideas were correct, his basic formulation needed some further work. The theory has now reached such a level of sophistication and acceptance that it is almost invariably included in discussions of the Big Bang. This addition was termed 'inflation'.

5.9   INFLATIONARY COSMOLOGY

The inflationary epoch is theorised to have taken place at the end of the GUT era when the Universe reached an age of 10^{-35} s. During discussions about the epoch of grand unification we have stressed how all the forces, apart from gravity, were indistinguishable. This is said to be a period of high symmetry and, in understanding inflation, we will draw the analogy that this era of the Universe can be represented by a lake of liquid water. The water in its liquid form is the same everywhere and, because there are no preferred locations or directions, the water is said to be in a symmetrical state. If the temperature of the lake were to fall, it would start to freeze. In the process of freezing, water molecules take up positions in a crystalline lattice, and it is no longer true that there are no preferred locations. Thus, in ice the liquid-water symmetry is broken. This change of state is analogous to events in the early Universe, when the temperature passed the critical value of 1×10^{28} K. The Universe changed state because the GUT symmetry broke down. No longer did the strong nuclear force act like the electroweak force.

When the water in the lake freezes it emits energy in the form of latent heat. This heat must be completely expended before the temperature of the ice can fall below $0°$ C. In our scenario for inflation, energy is emitted by breaking of the GUT symmetry, and this holds the Universe at a constant energy density for a short period of time. Maintaining the Universe at a constant density allows it to enter a period of exponential expansion – inflation.

This conclusion can be easily proved by recourse to the Friedmann equation. If the scale factor R(t) is differentiated with respect to time, the expansion rate of the Universe is determined. If the units of scale factor are appropriately chosen, thus formed is the equation

$$H(t) = \frac{R(t)}{R(t)} \tag{5.31}$$

If this is substituted into the Friedmann equation (5.29) it can be shown that the rate of universal expansion is proportional to the scale factor multiplied by the square root of the density. Since the density itself is proportional to the inverse cube of the scale factor,

$$R(t) \propto R^{-1/2} \tag{5.32}$$

In the inflationary epoch the density remains constant, because the energy emitted by the phase transition can be converted into matter by equation (1.2). This affects the expansion rate of the Universe, and it becomes

$$R(t) \propto R \tag{5.33}$$

If these two equations are plotted on a graph it can be easily demonstrated that the constant density of the inflationary epoch leads to an exponential increase in the size of the Universe; that is, it undergoes a period in which it continually doubles its size in a constant time (see Figure 5.9). At the beginning of the inflationary epoch, the characteristic time in which the Universe doubled its size was about 10^{-34} s. Even though inflation ended when the Universe reached an age of about 10^{-32} s it still had time to inflate by a factor of around 10^{50}!

This would solve the flatness problem because, for all but the most extreme geometries of spacetime, the stretching caused by inflation would flatten the surface (see Figure 5.10). By vastly inflating tiny areas of the Universe the horizon problem can also be solved.

At 10^{-34} s the Universe's horizon was 10^{-34} light seconds. If we were to extrapolate backwards from our present observed horizon, using equation (5.30), we would find that although small, the size of the Universe at 10^{-34} s would still be about 15 cm. The horizon distance represents the maximum scale upon which we can expect the Universe to be in thermal equilibrium and possess the same physical conditions.

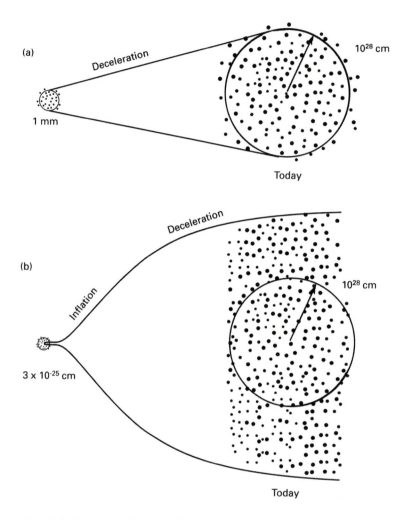

Figure 5.9. Inflationary and non-inflationary expansion. (a) In non-inflationary cosmology, today's observable Universe expanded from a region about 1 mm across at 10^{-35} s. Even though this is very small, it is still much larger than the horizon distance of the Universe at that time. (b) In an inflationary Universe, space is expanded during the period of inflation to become much larger than the horizon distance of the Universe. As a consequence, the Universe is much larger than we are able to observe. (Adapted from Silk, J., *A Short History of the Universe*, W.H. Freeman, 1994.)

Beyond the horizon distances, these disparate regions of the cosmos have had no time to communicate, and could be very different from each other. During inflation, every point within the Universe grew exponentially larger by a factor of 10^{50}. Inflation then stopped at 10^{-32} s, by which time the bubble of thermal equilibrium had been increased in size from 10^{-34} light seconds to 10^{16} light seconds (3×10^8 light

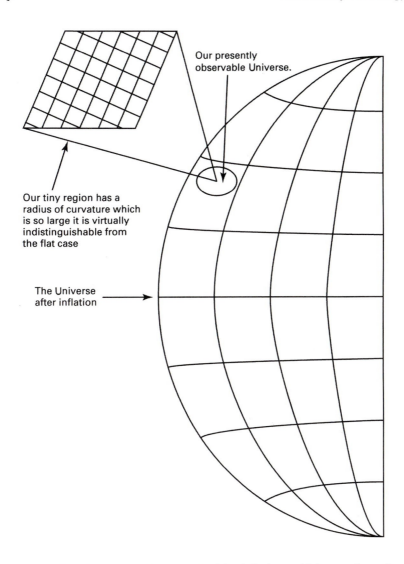

Our presently
observable Universe.

Our tiny region has a
radius of curvature which
is so large it is virtually
indistinguishable from
the flat case

The Universe
after inflation

Figure 5.10. The flatness problem is solved by inflation, which postulates that our Universe is a tiny region on the surface of a much larger curved surface.

years). This enormous expansion filled our Universe with energy which was all at the same temperature – and it solves the horizon problem.

In addition to these problems, it was mentioned earlier that the GUT phase transition could lead to topological defects in the spacetime continuum such as magnetic monopoles and cosmic strings. Although predicted by theory, these defects have yet to be found, although they have well-defined observational characteristics.

Inflation provides a reason why these objects have not to be found: as the Universe inflates, most of these newly-created objects will be pushed far beyond our cosmic horizon.

Finally, the energy released by the inflation signifies a fundamental change in the vacuum of space, and allows matter to form. In the coming chapters we shall refer to inflation several times, and indicate how it has helped shape our ideas – especially our ideas about galaxy formation and the cosmic microwave background.

6

The cosmological distance ladder

The previous chapters discussed the way in which the Universe has been observed to be expanding. The measurement of the scale of the Universe is important, and in order to calculate such fundamental quantities such as its expansion rate, its size, and therefore its age, it is necessary to acquire thorough knowledge of the distances to the objects which populate the cosmos. In essence this appears to be simple, and yet gauging cosmological distances is far from easy. Many methods abound, but in the broadest possible terms it is possible to divide the methods into two different types.

6.1 STANDARD CANDLES

A standard candle is any celestial object whose distance can be determined by a study of its brightness. The estimate is referred to as the luminosity distance. As this term suggests, the distance calculated by this method is based upon some kind of estimate of the true brightness of the object. Any deviation from this standard brightness is then attributed to the object's distance. This method relies on the inverse square law for the intensity of propagated light. Thus, if the distance of an object, d, trebles, then the observed flux, F, is reduced to one ninth of its value:

$$F \propto \frac{1}{d^2} \qquad (6.1)$$

If the luminosity, L, of the object is known precisely, then equation (6.1) can be used directly to determine a luminosity distance:

$$d = \sqrt{\frac{L}{4\pi F}} \qquad (6.2)$$

Often the precise luminosity is not known, and so other methods have to be found in order to estimate an object's distance. Astronomers use the magnitude system to reference a celestial object's brightness. Apparent magnitude, m, is the brightness of a star (or other body) as it appears in the sky; but it does not account for the distance of the object under scrutiny. It is very likely that a faint, close-by object will appear to be much brighter than a distant but intrinsically bright object. A better system for measuring brightness is absolute magnitude, M – the apparent magnitude of a celestial object if it were at a distance of 10 parsecs. Absolute magnitude therefore provides a direct comparison between the brightness of all known celestial objects; and a combination of the two systems can lead astronomers to the distance modulus.

The magnitude system was introduced during the period of the ancient Greek civilisation. In the late third century BC, the Greek astronomer Hipparchus classified stars according to their brightness. He divided them into six different categories, and classed the brightest as being of first magnitude and the faintest as being of sixth magnitude. With the invention of the telescope, however, this system had to be revised and extended. Although William Herschel originally undertook this endeavour, our modern magnitude system is based upon the work carried out by the English astronomer, Norman Pogson, during the 1850s. To mimic the eye's response, Pogson formulated a mathematical equation which used logarithms. In studying the classification scheme introduced by Hipparchus, he discovered that the relationship between each of the magnitudes was approximated by a factor of 2.5. He therefore devised an equation to imitate this relationship by comparing two apparent magnitudes, m_1 and m_2, and the flux of radiant energy, which could be measured from the celestial objects at the surface of the Earth, F_1 and F_2.

$$m_1 - m_2 = -2.5 \log \left(\frac{F_1}{F_2} \right) \tag{6.3}$$

This has become known as Pogson's equation, and is the starting point for our derivation of the distance modulus. It was the introduction of a rigid mathematical formulation for magnitude which resulted in some objects possessing magnitudes with a negative value. Thus, an object with a magnitude of –1 is 2.5 times brighter than an object with a magnitude of 0. This object is, in turn, 2.5 times brighter than an object of magnitude 1.

Instead of the flux of energy received at the Earth's surface, if the actual energy released per unit area (the luminosity) of each of the celestial bodies is known, then Pogson's equation instead produces absolute magnitudes:

$$M_1 - M_2 = -2.5 \log \left(\frac{L_1}{L_2} \right) \tag{6.4}$$

The luminosity of the object and its measured flux are linked by the inverse square law (equation (6.1)), so that

$$F = \frac{L}{4\pi d^2} \tag{6.5}$$

If this is to be used to determine absolute magnitude, then the distance, d, is fixed at 10 parsecs. To obtain the distance modulus, imagine observing an object of a known celestial type which has an absolute magnitude of M. Observations of the flux for this object can derive an apparent magnitude by using equation (6.3) in conjunction with a standard reference star. Thus an apparent and an absolute magnitude are now known for the object. These magnitudes and their associated fluxes can be substituted into equation (6.4):

$$m - M = -2.5\log\left(\frac{L4\pi(10)^2}{L4\pi d^2}\right) \tag{6.6}$$

This equation simplifies in a few stages.

$$m - M = -5 + 5\log d \tag{6.7}$$

This is the distance modulus equation, which can be used to determine a luminosity distance, in parsecs, providing that the absolute magnitude of the object under scrutiny is known. It assumes that the only reason for the object to diminish in brightness is the inverse square law; but this is not strictly true. Interstellar dust particles can also dim celestial objects by scattering and absorbing photons of light. This interstellar extinction can be estimated, and can then be included in the distance modulus expression:

$$m - M = -5 + 5\log d + Ad \tag{6.8}$$

A is a constant extinction factor, and is estimated to be about 0.002 mag/pc for a line of sight in the plane of our Galaxy. The distance modulus and Pogson's equation are repeatedly encountered when studying the cosmological distance ladder because, in various guises, they link the observable properties of celestial objects – such as apparent magnitude and fluxes – to distance.

6.2   STANDARD RULERS

A standard ruler is a celestial object with a linear diameter, l, of which we can be reasonably certain. Thus, if the angular diameter of a distant object, θ, is measured, trigonometry can then be used to construct a right-angled triangle, and the distance, d, can be easily calculated. In its simplest form, this so-called 'diameter distance' can be calculated using the equation

$$d = \frac{l}{\theta} \tag{6.9}$$

By modifying the standard ruler approach, the most accurate method of distance determination is available. It is known as trigonometric parallax, and is possible only because the diameter of the Earth's orbit is well known. As the Earth orbits the Sun, it causes us to view the Universe from different vantage points throughout the year. This difference is at a maximum every six months because the Earth is on different sides of its orbit. In principle, trigonometric parallax involves an angular measurement of a star's shift in position relative to the background stars, which can then be used to construct a triangle from which its distance can be calculated (see Figure 6.1).

In 1727, Englishman James Bradley attempted to measure the parallax of the star

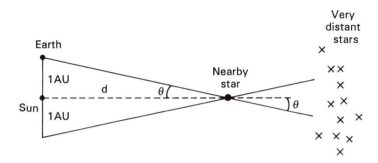

Figure 6.1. Trigonometric parallax is the best method for determining stellar distances. Unfortunately it is very limited in range but the European Space Agency's Hipparcos satellite has significantly improved this situation.

γ Draconis. Over a six-month period he did indeed measure a shift in its position relative to the background of stars, but it was far too great to be due to parallax. Bradley had instead discovered the phenomenon of aberration, which occurs because light travels at a finite speed, and because the Earth is a moving platform from which to make observations.

Parallax was finally measured over a century later by the German mathematician Friedrich Bessel, who studied the double star 61 Cygni. This star was found to have a tiny parallax which indicated a distance of some 6×10^{13} miles (1×10^{14} km).

It is parallax which provides astronomers with their unit of distance: the parsec (parallax arcsecond), which is defined as the distance a star would be from Earth if it were to display a parallax of one second of arc. This equates to a distance of approximately 3.26 light years. In reality, all parallaxes are smaller than one arcsecond because stars in our celestial neighbourhood are spread so far and wide. The nearest star to us, α Centauri, possesses a parallax of 0.75 arcseconds.

This illustrates the one problem with parallax: it is measurable only for the nearest stars to our Solar System. The displacements are so small that the calculated distances rapidly become inaccurate; and so until recently only about 1,000 stars had had their distances measured directly by this method. However, a dramatic advance in parallax measurement came with the European Space Agency's spaceprobe Hipparcos – the HIgh Precision PARallax COllecting Satellite. The primary mission of Hipparcos was to measure positions, parallaxes and proper motions of 120,000 stars to an accuracy of 2–4 milliarcseconds. It succeeded, and the gargantuan catalogue which was subsequently published provides the strongest foundation that mankind has ever had upon which to base the cosmological distance scale.

6.3   PRIMARY INDICATORS

A primary distance indicator is a celestial object which has been calibrated by observations of similar objects within the Milky Way. The distance will preferably have been calculated by parallax measurements, but other observational or theoretical factors will often have to be considered. Having established the observational properties of the primary indicator in our own Galaxy it can then be compared to similar objects in other galaxies, either by using it as a standard candle or a standard ruler.

6.4   CEPHEID VARIABLE STARS

The very best of the primary indicators are the Cepheid variable stars, named after the prototype of this class, δ Cephei. They are yellow stars, and occupy a region of the Hertzsprung–Russell diagram just above the main sequence. They represent the next stage in a star's evolution after it has passed through the red giant stage. Many stars are thought to enter this stage of existence when helium begins to fuse into carbon in their cores. The breakthrough in using these stars as distance markers came at the beginning of this century, and was made by the astronomer Henrietta Leavitt, working at Harvard College. She obtained photographic plates of the Magellanic Clouds which, in those days, were known only as being incredibly dense star clusters. She identified many variable stars on the photographs, and began to notice a correlation between brightness and the time taken to pulsate – a quantity known as the period. The brighter the star, the longer it took to pulsate. She then compared the shape of these variable stars' light-curves with the light-curves of variable stars in more immediate surroundings. She discovered that the curves were exactly the same as the Cepheid variable stars, but the brightnesses were different. This she attributed to the variable stars in the Magellanic Clouds being much further away. The distances to nearby Cepheids had yet to be determined, however, because none were in the range of parallax measurements. The final piece in the puzzle was added by Harlow Shapley, who realised that the variations in brightness must be due to variations in the size of the star during pulsation. He used this theory to calculate the size of the individual Cepheids and their luminosities. As soon as this information was in place, luminosity distances became easy to determine. It was the discovery of Cepheid variable stars in the Andromeda galaxy which allowed Edwin Hubble to conclude that it was a galaxy external to our own and, by extension to other galaxies, propose the redshift–distance law (see Figure 6.2).

The relationship between the average absolute magnitude of a Cepheid variable star, M_v, and its period of pulsation, P, can be quantitatively expressed by the equation

$$M_v \cong -2.8 \log P \qquad\qquad (6.10)$$

which is strictly valid only for younger, Population I Cepheid variables, known as classical Cepheids. Older, metal-deficient Population II Cepheids are referred to as W Virginis stars, and these follow the slightly modified period–absolute magnitude relationship

$$\overline{M}_v \cong -1.9 -2.8 \log P \qquad\qquad (6.11)$$

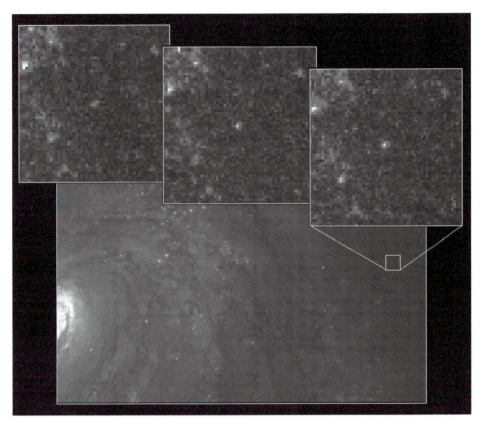

Figure 6.2. A Cepheid variable in M100. The Hubble Space Telescope has identified Cepheid variables in the spiral galaxy M100, a galaxy which is thought to lie in the Virgo cluster of galaxies. The top panels illustrate the variation in light of the Cepheids as observed by the HST over a period of time. (Photograph reproduced courtesy of Dr Wendy L. Freedman, Observatories of the Carnegie Institution of Washington, and NASA.)

A third type of related variable star is the RR Lyrae star. These can be thought of as short-period Cepheids. They pulsate with periods between 0.25 day and 1 day, and have a mean absolute magnitude of about 0.8. Both classes of Cepheid have absolute magnitudes in the range −1 to −7, and pulsation periods in the range 2–100 days (see Figure 6.3). RR Lyrae stars seem to be found predominantly within globular clusters, and at one time were referred to as cluster variables.

The limit to using Cepheid variables exclusively to gauge distance is determined only by technology. However, as bigger and better telescopes with adaptive optics systems are being built, so the range increases at which Cepheids can be discerned.

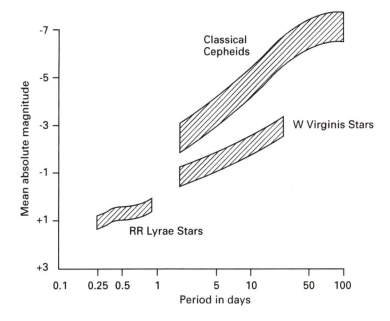

Figure 6.3. The period–luminosity relationship for Cepheid variables provides one of the most accurate primary distance determination methods presently available.

6.5   OTHER PRIMARY INDICATORS

A nova is thought to be a close binary star system in which one of the components is a white dwarf star. The other component is a giant star which has grown so large that the gravity of the white dwarf can strip off its outer layers of hydrogen. This gas spirals down onto the white dwarf and builds up on its surface. When sufficient material builds up, the gravity of the white dwarf can cause thermonuclear detonation of the hydrogen, which results in a huge brightening of the system, observable as a nova.

On theoretical grounds it is thought that 10–40 novae occur within the Milky Way each year. Only two or three of these are observed from Earth, however, because the others are obscured by the vast quantities of interstellar dust and gas which are located within our Galaxy. A nova appears to be a fairly standard outburst and can be characterised by a generic light-curve. The principal features appear to be a very rapid rise in brightness by 8–10 magnitudes, followed, after a short pause, by another rise of a few magnitudes, after which the star begins to fade (see Figure 6.4).

Typically, the peak absolute magnitude of a nova is −8, but a reasonably precise empirical relationship exists between the peak absolute magnitude, M_v, and the time taken for the nova to decline by three magnitudes, t:

$$M_v \cong -11 + 2 \log t \qquad\qquad (6.12)$$

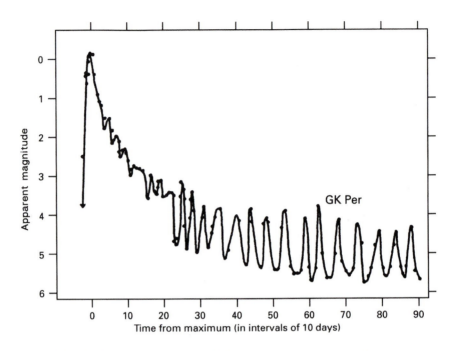

Figure 6.4. The light-curve of nova GK Persei. Novae generally follow very similar patterns of outbursts. The typical light-curve shown here can be used, in conjunction with a statistical analysis of many other nova light curves, to calculate a luminosity distance.

This relationship allows novae to be used as standard candles and have their distance moduli calculated.

Other primary indicators include the matching of the spectral type of stars in distant galaxies to those in our own Galaxy. Sampling a range of stars for their spectral types can allow the construction of a Hertzsprung–Russell diagram for the distant galaxy. The main sequence will be present, but at dimmer magnitudes than the corresponding diagram for our Galaxy. By analysing this difference, a luminosity distance to the galaxy can be determined.

6.6   SECONDARY INDICATORS

A secondary distance indicator is an object whose distance is determined by the calibration of a primary indicator. For example: if the distance to a globular cluster is determined by examination of an RR Lyrae star – a primary indicator – the whole globular cluster can be used as a secondary distance indicator. This would involve comparing its properties with those of a globular cluster around an even more distant galaxy.

Other secondary distance indicators are H II regions – emission nebulae – which are produced by star-formation regions, such as the Orion nebula. The luminosity of the Hα emission line at 656 nm can be used to transform these objects into standard candles; or their diameters can be used as standard rulers.

6.7 SUPERNOVAE

Supernovae are celestial catastrophes, so bright that they are visible on all celestial scales – from close by to (almost) the limits of the observable Universe. They are, therefore, potentially very important distance indicators (see Figure 6.5). Two types of supernova have been identified. Type I supernova are thought to occur in stellar systems of similar configuration to those which cause novae. In these cases, the accreting white dwarf collects so much material that it undergoes gravitational collapse. This produces an almighty stellar explosion, ripping the remains of the white dwarf to pieces, which we observe as the supernova. These supernovae can be subdivided into three types. Of these, only supernovae Type Ia supernovae are really useful for distance determination, because the other two types, b and c, display irregularities in their light-curves and spectra, and are also fainter and rarer.

The reason that a Type Ia supernova has a sharply defined absolute magnitude peak is that the mass limit for gravitational collapse is sharply defined: 1.4 solar masses. Beyond this limit, the atomic structure of matter breaks down, and objects become compacted balls of neutrons. As this collapse proceeds, nuclear reactions occur because of the vast amounts of energy liberated by the process. Thus, every Type Ia supernova has been precipitated by the collapse of a white dwarf of mass 1.4 times that of the Sun.

Typically, a Type Ia supernova reaches a peak absolute magnitude of –19, the rise to maximum proceeding at a rate of 0.25–0.5 magnitudes per day. After a week or so at maximum brightness, the decline in brightness begins. This initially occurs at a rate of one magnitude per week, but soon slows down to just one magnitude every ten weeks. As well as being excellent standard candles, supernovae can be used as standard rulers. The explosion produces an expanding envelope of luminous, gaseous matter, which can be studied spectroscopically to determine its expansion velocity from the Doppler shift of its spectral lines. The angular expansion of the supernova remnant can also be measured, and trigonometry employed, to determine the distance.

A result from the Hubble Space Telescope in 1996 has boosted confidence in the use of Type Ia supernovae as secondary indicators of distance. A team of astronomers, led by Allan Sandage of the Carnegie Observatories, used the superior resolution of the HST to identify a number of Cepheid variables in the galaxy NGC 4639. They chose this particular galaxy – located in the Virgo cluster – because in 1990 a supernova of Type Ia was observed there. The Cepheid variables yielded a distance of some 78 million light years, which equates with the distance calculated from the supernova.

Supernovae of Type Ia can also be used to study the expansion rate of the Universe as a function of time, because they are so bright and can be seen over

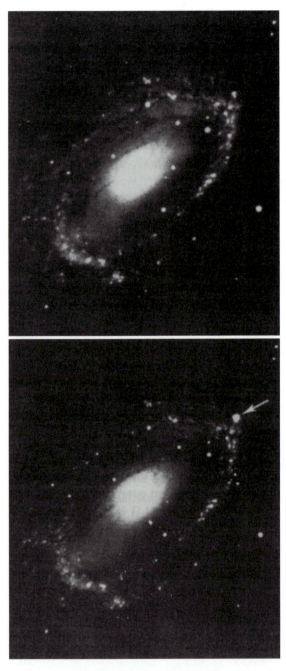

Figure 6.5. Supernova in NGC 4725. In the lower photograph, taken on 2 January 1941, the supernova (arrowed) is clearly visible, while in the upper photograph, taken earlier, on 10 May 1940, it is invisible because it has yet to explode. (Photograph reproduced courtesy of Mount Palomar Observatory.)

enormous distances. Just such a study – the Supernova Cosmology Project – was undertaken by Professor Richard Ellis, of the Cambridge Institute of Astronomy, and his collaborators. They first began working on this approach in the 1980s, and by 1988 had located a supernova in a cluster of galaxies at a redshift of $z = 0.31$. The project was feasible only if a powerful survey technique could be found for observing many galaxies and finding as many supernovae as possible. Saul Perlmutter and Carl Pennypacker, both of Lawrence Berkeley Laboratory, California, joined the project by offering the use of their panoramic CCD camera, which was capable of surveying many thousands of faint galaxies in just a few nights of observation. The observing runs were always timed to coincide with the new Moon, so that the sky would be as dark as possible, allowing the faint galaxies to be observed. Typically, about ten supernovae were discovered on each observing run. These were then observed repeatedly so that their light-curves could be plotted. Their spectra were also taken so that their redshifts could be calculated, and also to ascertain whether they were of the correct supernova type – Type Ia.

Since Type Ia supernova progenitors are old binary systems containing a white dwarf, they are usually found far away from the dusty areas where new stars are formed. The light from the explosion is released into intergalactic space, and is extinguished mostly as a result of its inverse square law propagation throughout space. By comparing the apparent magnitude of the supernova with its theoretical absolute magnitude, the distance modulus (equation (6.7)) can be calculated.

Since distance is correlated to redshift via the changing expansion rate of the Universe, a graph of the apparent magnitude against redshift for a number of such supernovae can highlight the way in which the expansion rate of the Universe has changed with time. The initial results of this approach were so successful that the team began using the Hubble Space Telescope to extend their observations, and discovered supernovae at extreme distances. The analysis of these supernovae has led to a surprising conclusion that is currently fascinating cosmologists around the world. Far from the traditionally expected decelerating Universe, the supernova cosmology project hints that the Universe is accelerating. (This fascinating aspect of the work will be explored in more detail in section 11.9.)

The second type of supernova occurs when a massive stars exhausts its supply of nuclear fuel and collapses. In the heart of a star greater than eight times the mass of the Sun, the nuclear furnace is so intense that all of the chemical elements up to iron can be synthesized. In a star with lower mass, this process largely finishes once helium has been transmuted into carbon. Iron is a very stable element which is very difficult to fuse, and it builds up as an inert core in the centre of the star. When it reaches the Chandrasekhar limit, the atomic structure of the iron breaks down and the star collapses. As with a Type I supernova, this causes a cataclysmic explosion. Although a Type II supernova can be used to calculate distance, there is a much greater spread in the peak luminosity. This is because the mass of the progenitor star is usually unknown, and hence estimation is necessary.

6.8 THE TULLY–FISHER RELATIONSHIP

A rather novel approach to distance determination links two observable properties of the galaxy in question. The first property is the apparent magnitude of the galaxy. This is obviously dependent upon the distance to the galaxy, and it can be measured at optical wavelengths or in the infrared. (The latter is often better, because the extinction in the light caused by the intergalactic medium is not so great.) Secondly, a method of calculating the galaxy's intrinsic luminosity must be determined. This is the function of the second observation. Using radio telescopes, the galaxy is observed, with the intention of detecting its interstellar medium, by measuring the 21-cm emission line of molecular hydrogen. As the galaxy rotates, so the hydrogen clouds' emissions are Doppler-shifted. This creates a broadening of the line, because the individual clouds move in many different directions. Thus, the aim of this measurement is to determine just how broad the line has become because of the rotation of the galaxy. The galaxy's mass can then be estimated, and its intrinsic luminosity can be calculated. Some measure of the galaxy's inclination to our line of sight must also be carried out so that a correction factor can be applied. Once the value for the luminosity is calculated, the luminosity distance of the galaxy can be determined.

6.9 TERTIARY INDICATORS AND GRAVITATIONAL LENSES

The final type of distance indicator relies on the comparison of the global properties of galaxies. One galaxy in a cluster will have had its distance determined by a secondary indicator: for example, the largest galaxy in a cluster may be used as a standard ruler. This assumes that the largest galaxy in another more distant cluster will be of equal extent, and thus both can be directly compared. As well as using whole galaxies as standard rulers, the brightest galaxies in clusters can be used as standard candles. Once again, similar assumptions are needed to facilitate the use of this method.

Another tertiary indicator is the luminosity classification of spiral galaxies by the appearance of their spiral arms. A class I object will have the brightest, most regular and well-defined structure to its spiral arms, whereas a class V object will be completely the opposite. As we shall see later in this chapter, the applicability of some spiral galaxies as standard candles has been called into question.

Finally, one of the most promising new distance determination methods relies on the monitoring of gravitational lenses. These are fascinating objects in which the image of a distant quasar has been distorted because it has interacted with the gravitational field of an intervening galaxy (see Figure 6.6). According to quantum theory, a photon travelling at the speed of light will display the property of momentum. Thus, in some way or another it behaves as if it possesses a mass, and should therefore be influenced by a gravitational field. This should not really be a surprise, because of Einstein's equation (1.2), which he derived as part of his special theory of relativity. During the total solar eclipse of 1919, the bending of light in the

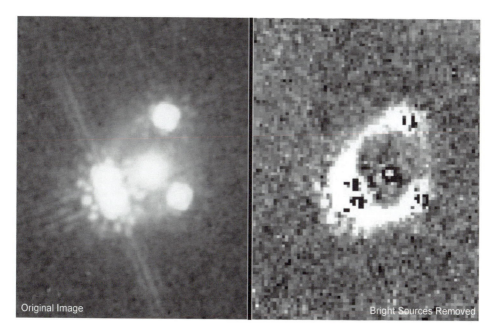

Figure 6.6. Careful processing of this image has revealed that the gravitational lens has produced an almost complete Einstein ring. (Photograph reproduced courtesy Christopher D. Impey (University of Arizona) and NASA.)

presence of a gravitational field was proven to be exactly as Einstein's general relativity predicted. During this event, the precise positions of stars close to the Sun were measured, and it was found that they had indeed shifted their positions by the amount predicted if the Sun's gravitational field was altering the paths of their photons.

There are several examples of double quasars in the night sky which, as well as being physically close together, also show exactly the same spectrum. The first of these was found in 1979 by a team of astronomers led by D. Walsh. The reason for multiple images of the same quasar is that between the quasar and ourselves there is an intervening galaxy. In some cases the galaxy is so faint that it cannot even be seen; its existence is intimated by its production of a multiple quasar image. The number of images and the relative brightness between them is dependent upon the precision of the alignment between ourselves (the observers), the lensing galaxy and the distant quasar. If all three are in perfect alignment, then the quasar image will be distorted into a ring, known as an Einstein ring. If the alignment is not perfect, a number of distorted images will surround the lensing galaxy (see Figure 6.7).

To understand why multiple images are formed, the quasar should be considered as emitting light in spherical wavefronts which propagate radially through the Universe. At a certain distance on its journey through space, part of the wavefront enters the gravitational field of a galaxy. In a way analogous to waves in water slowing down when they encounter shallow water, so the photons slow down in the presence of a gravitational field (as viewed by outside observers). In the case of the

Figure 6.7. This image of the rich galaxy cluster Abell 2218 is a spectacular example of gravitational lensing. The gravitational field of this massive compact cluster deflects light-rays passing through it – a process which magnifies, brightens and distorts images of objects that lie far beyond the lensing cluster. The thin arcs, spread across the picture like a spider's web, are the distorted images of very distant galaxies 5–10 times more distant than the cluster. (Photograph reproduced courtesy of W. Crouch, University of New South Wales, R. Ellis, Cambridge University and NASA/STScI.)

quasar wavefront moving through the galaxy, the closer to the galactic centre it passes, the deeper into the gravitational well it will travel and the slower it will become. Thus, the wavefront will become deformed, with those parts which have travelled through the extremities of the galaxy having been deflected from their original paths.

As the wavefront leaves the influence of the galaxy, its propogation is no longer totally radial. Instead, parts of it have a component of motion which has been caused by the gravitational field of the galaxy, which causes segments of the wavefront to fold over. When the wavefront is observed from the Earth, and as each segment hits our detectors, a different image of the quasar is perceived. The images lie in slightly different directions, because the individual wavefronts hit us at slightly different angles, and our brains interpret what we have just seen to lie in a direction which is perpendicular to the wavefront (see Figure 6.8). The folding of the wavefront obviously implies that different sections arrive at slightly different times. If a change in the quasar can be monitored, and the time taken for each image to respond to this change, the distance between the wavefront segments can be calculated. This can then be combined with other estimates of the lensing system's properties in order to determine distances to both the galaxy and the lensed quasar.

Although this method still requires an assumption about certain properties, it is not a standard candle or standard ruler technique. It is therefore a potentially important tool for corroborating a distance obtained by the standard methods.

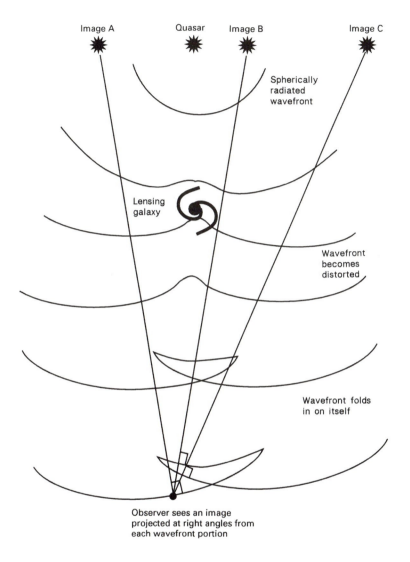

Figure 6.8. The principle of a gravitational lens system, showing the origin of the multiple images caused by the distortion in the quasar's wave front.

6.10   PROBLEMS IN DISTANCE DETERMINATION AND THE MALMQUIST BIAS

There are many problems to consider when estimating the distance to a celestial object. The zero point error is a mistake made in the determination of the distance to a primary indicator. This error is then propagated up the distance ladder through use of secondary and then tertiary indicators.

The extinction suffered by a standard candle in a galaxy will also cause inaccuracies in the calculated distance. Extinction occurs because of the interstellar matter within the galaxy containing the object, and is also caused by intergalactic matter and then by interstellar matter within the Milky Way. Those objects which are viewed at high Galactic latitudes (out of the plane of the Milky Way) will suffer less from this problem, although they will still be subject to extinction by intergalactic matter and by dust within their host galaxies.

Even if extinction were non-existent, there would still not be a perfect standard candle. Although this chapter quotes several peak absolute magnitudes for various celestial objects, these are only average magnitudes. Each and every one of them is capable of possessing a value for its absolute magnitude which can vary from this mean value. This range of values is known as the dispersion. In more and more distant galaxies it will be easier to detect standard candles which have luminosities towards the high end of the intrinsic dispersion. This, obviously, causes errors in the distance determination. The deficiency of detecting only the brightest objects in distant galaxies – the Malmquist bias – can lead to an underestimate of the distance (see Figure 6.9).

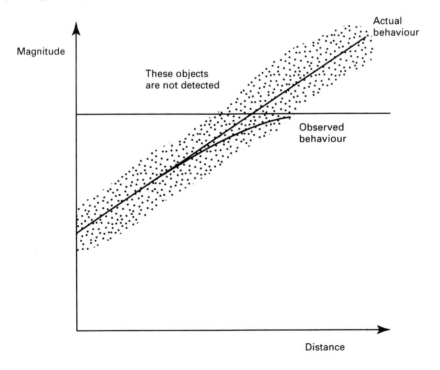

Figure 6.9. The Malmquist bias is a common distortion encountered when using standard candles. At larger and larger distances, the number of observed standard candles reduces, because the fainter population members are undetectable. This introduces a distortion in the calculated luminosity distances.

6.11   CALIBRATING THE REDSHIFT SCALE

As we have seen in Chapter 1, all galaxies and other extragalactic objects, apart from those in the Local Group, are subjected to the expansion of the Universe. This causes a redshift to be imparted upon the spectra of these objects. It has been established that, in some way, the distance of a galaxy is linked to its redshift. The Hubble law (equation (1.4)) proposes that this linking is simply a linear relationship. Thus, a galaxy which exhibits twice the redshift of another lies at twice the distance. In order to calibrate the relationship, a constant is needed: the Hubble constant. The value of this is calculated by measuring the distance to a galaxy by one of the methods described above, and then equating this with its redshift measurement. Although this sounds very easy in principle, in practice it is far from being so. Errors and approximations have crept into the cosmological distance ladder, so that the value of the Hubble constant is currently uncertain by a factor of two. The general consensus places it around 50–100 km/s/Mpc.

It is interesting to note that when determining the Hubble constant, astronomers seem to fall into one of two camps. This rivalry began in 1976, at an international colloquium on 'Redshift and the Expansion of the Universe'. Gerard de Vaucouleurs presented a value for the Hubble constant of approximately 100 km/s/Mpc, whilst G.A.Tammann and Allan Sandage presented, from result of their combined study, approximately 50 km/s/Mpc. Interestingly enough, the passage of 20 years has done little to narrow the gap between these two estimates. They have become known as the 'short-scale' and the 'long-scale' respectively.

The reason for these names is explained as follows. The Hubble constant, with its peculiar measure of km/s/Mpc, will actually simplify to a value which possesses the dimensions of inverse time (s^{-1}). Thus, if the reciprocal of the Hubble constant is taken, a time is obtained. As will be shown in Chapter 11, this time is actually an upper estimate for the age of the Universe, and is called the Hubble time. A Hubble constant of 50km/s/Mpc produces a Hubble time of approximately 20 billion years. Doubling the Hubble constant halves this time to 10 billion years. Reference to the Hubble time therefore leads to high estimates of the Hubble constant being termed 'short-scale'. Similarly, low estimates are collectively termed 'long-scale'.

An interesting point to note about those rival astronomers who estimate the Hubble constant is that, in many instances, they analyse exactly the same observational data but still arrive at different estimates. The only real difference has been in the assumptions they have made about the validity of the data, and the way in which they have estimated its inherent errors. Short-scale estimates present a real problem for cosmology, because their associated Hubble time is less than the estimated age of the oldest stars in globular clusters.

Many astronomers feel comfortable using a value for the Hubble constant of about 75 km/s/Mpc. They often feel that because some working methods inherently provide high estimates of the Hubble constant and others provide low estimates, a statistical average between the two is necessary. Other astronomers, however, remain firmly rooted in either the long-scale or short-scale camps because of a deep conviction that the other method is incorrect.

The biggest problem in calibrating the redshift scale is that we are observing a moving object from a moving platform. Imagine a galaxy being scrutinised by an astronomer on the Earth. As well as the cosmological redshift imparted on the galaxy in question by the expansion of space, any motion of the Earth or the galaxy along the observer's line of sight will cause a Doppler shift which will either enhance or diminish the cosmological redshift.

Consider the Earth. Not only is it rotating, but it is also in orbit around the Sun. The Sun is in orbit around the centre of the Galaxy, and the Galaxy is in orbit around the centre of mass of the Local Group. The Local Group itself is being influenced to move by the gravity of the largest galaxy clusters around it. Unless the galaxy being studied is at a precise right angle to the direction in which the Earth is moving, some change in its redshift will be introduced. Astronomers, therefore, are hardly in an ideal situation (see Figure 6.10). Luckily, the cosmic microwave background radiation comes to our aid. The combination of all the Earth's motions through space is revealed in our observations of the microwave background as a dipolar anisotropy caused by the Doppler effect. In other words, the frequency of the microwaves is increased slightly in the overall direction of the Earth's motion. Thus, an estimate of the Earth's true velocity – with reference to the cosmic microwave background – can be calculated.

It is far more difficult to determine the peculiar motion of the galaxy being studied. Not all regions of space are expanding in accordance with the Hubble flow; certain places in our Universe contain so much matter that they create a gravitational field which is sufficiently strong to resist the expansion of space. These places are known as clusters – slightly smaller galaxy groups (see Figure 6.11). Within these associations, each of the individual galaxies is in orbit around the centre of mass. Near the centre of the cluster – where galaxies are deep in the cluster's gravitational well – the velocities possessed by the individual galaxies can be large, and are often sufficient to significantly alter their redshifts. A perfectly spherical cluster will actually look like a skewed ellipse if the redshifts of its constituent galaxies are plotted on a graph.

Smaller clusters are moved by the gravity of larger ones – a phenomenon known as streaming. These often large motions have been discovered only in recent years, and present a significant problem for cosmologists. In Chapter 4 the scale upon which the cosmological principle could be applied was stated to be hundreds of millions of light years. On a smaller scale, the distribution of matter in the Universe is uneven and lumpy. Thus, less dense regions will be attracted by the gravity of overly dense regions, and streaming will occur.

Only large-scale mapping projects such as those which are currently using multiple fibre-optic spectrometers – for example, the 2-degree Field at the Anglo-Australian Telescope, or the SLOAN Digital Sky Survey – will make headway in mapping the three-dimensional distribution of galaxies throughout the Universe. Once a map of the cosmos has been constructed, large-scale streaming motions will be better understood.

The Hubble Space Telescope has been involved in a most interesting project concerned with the cosmological distance ladder. A key project of the telescope's

Figure 6.10. Compact galaxy group Hickson 40. Galaxy groups in the Hickson catalogue contain galaxies so closely associated that they are gravitationally interacting. (Photograph reproduced courtesy of National Astronomical Observatory of Japan.)

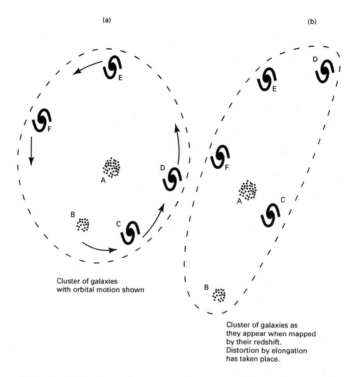

(a)

(b)

Cluster of galaxies
with orbital motion shown

Cluster of galaxies as
they appear when mapped
by their redshift.
Distortion by elongation
has taken place.

Figure 6.11. Redshift distortions occur in galaxy clusters because there is no easy method of separating the cosmological redshift from the individual Doppler shifts caused by the galaxies' motions through the cluster. Because of this effect a spherical cluster is elongated, which causes the features known as 'fingers of God' that appear to point at Earth on redshift maps of the Universe.

first five years in orbit was the task of refining the accuracy of the Hubble constant to within 10% of its actual value. To this end, an international team of more than 20 astronomers surveyed galaxies in both the Virgo cluster and the Fornax cluster. Both clusters are at approximately the same distance, but in different directions. The Virgo cluster is very large, and hence the redshift distortions suffered by its constituents are potentially large. The Fornax cluster, however, is much more compact, and should display redshifts which are in better agreement between the individual galaxies (see figure 6.12). The other advantage of studying the two clusters in parallel is that, since they are in different parts of the sky – hence different regions of the Universe – large-scale streaming motions affecting one may not necessarily affect the other.

The investigating team has been deriving values for the Hubble constant which fall in the range of 68–78 km/s/Mpc. However, before this chapter leaves the reader with the impression that a consensus is finally being reached, it is well worth noting that another, independent investigation undertaken by astronomers using the Hubble Space Telescope – and reported just two months before the above study – resulted in a derived value of just 57 km/s/Mpc. The controversy continues to rage.

Figure 6.12. NGC 1365 in Fornax. The Hubble Space Telescope has surveyed the Fornax cluster of galaxies for Cepheid variable stars. Although at a similar distance, the Fornax cluster is a tighter collection of galaxies than is the Virgo cluster, and so the determination of the Hubble constant may not be as susceptible to distortions from motion around the cluster's common centre of mass. NGC 1365 is a prominent Fornax cluster member, and was one of the galaxies studied by the HST team. This image of the galaxy was taken by the European Southern Observatory's Very Large Telescope in Chile. (Photograph reproduced courtesy of ESO.)

6.12 IS THE REDSHIFT COSMOLOGICAL?

Our current view of the cosmos depends crucially on one bold assumption: that the redshift displayed by distant celestial objects is caused predominantly by the expansion of the Universe. This chapter has already discussed the confusion caused by the motion of a galaxy through its parent cluster, which imparts an additional Doppler shift on its radiation. Some astronomers believe that there may be an as yet undiscovered mechanism which can cause a redshift. Although these astronomers are at present in a minority, that single fact alone cannot be used to dismiss their theories.

The question of whether or not the redshift is solely dependent upon the cosmological expansion of space rests with either proving or disproving the Hubble law (equation (1.4)). There is a mounting body of evidence against this simple linear relationship, which must be explained within the framework of the conventional theory, or it will eventually topple current beliefs. For some it has already done so, and they are busy working on alternatives to the expanding Universe model.

In an attempt to illustrate the problems which still face the Hubbleian proponents of an expanding Universe, this section will describe a few of the most well-known 'flies in the cosmic ointment'. It is worth mentioning that, although there are several astronomers who have worked on observations and theories which cast doubt upon the expanding Universe and Big Bang theories, it is Halton Arp's name which is most prominent. Arp is a tireless dissenter from the conventional viewpoint, but it must be stressed that this is not one-man crusade. Respected astronomers from all over the world have obtained data which are difficult to explain within the current framework.

The first example derives from within our own Local Group. Study of the Andromeda galaxy has shown that it is surrounded by a number of satellite galaxies (see Figure 6.13). Four have been identified: M32, M33, NGC 205 and NGC 185 – all of which are redshifted with respect to the Andromeda galaxy. It should be noted here that the Andromeda galaxy itself shows a blueshift, because it is being gravitationally attracted towards the Milky Way. Thus, when the companion galaxies are described as being redshifted with respect to it, it actually means that they are less blueshifted. If these companions are in orbit around the central galaxy, we would expect some to be blueshifted in relation to it. The only conventional explanation for the observation that they are not blueshifted is that the galaxies surrounding the Andromeda galaxy are all moving away from us in their respective orbits. This would be a large coincidence, but would nevertheless be acceptable. The same pattern of discrepant redshifts are observed to repeat, however, in other galaxy systems, notably for the companion galaxies of M81.

The leading advocate of this result, Halton Arp, has gone further. He has even included all galaxies smaller than the dominant one from the cluster in the analysis. His work on the Local Group and the M81 group certainly appears to show that the smaller galaxies in these groups display larger redshifts than the dominant galaxy of the group. Taken at face value, this result would seem to indicate that a proportion of the redshift of a galaxy is dependent upon its size. There is evidence, too, that the Hubble type may also influence a galaxy's redshift. The Hubble classification of spiral galaxies divides them into three sub-groups. They were classed according to the apparent size of their nuclei, and the tightness of the winding of their spiral arms. Of the three types, the Sc galaxies were those with the smallest nuclei and the most loosely wound arms. This class has since been further subdivided with the inclusion of Sc I galaxies, which are undeniably Sc in nature, but which have spiral arms which are narrower and better defined.

Figure 6.13. The Andromeda galaxy, M31 is the spiral galaxy closest to us. Studies of its satellite galaxies have shown that all of them possess a relatively larger redshift component than does the parent galaxy. This is curious, because the galaxies are assumed to be in orbit around M31 and should therefore display a spread of redshifts and blueshifts. (Photograph reproduced courtesy of Photolabs, Royal Observatory, Edinburgh.)

The problem with these galaxies is that, if their linear diameters are calculated, based upon their angular diameters and their redshift distances, they become larger and larger with increasing redshift. It would appear that if redshift is a totally reliable distance indicator, the Sc I galaxies at earlier epochs were bigger than the current examples. This problem can be so severe that, if the redshift distance of Sc I galaxy NGC 309 is correct, then it totally dwarfs the Sb galaxy M81 (see Figure 6.14). Conversely, based upon examples in the present-day Universe, the general rule is that Sb galaxies are larger than Sc I galaxies.

In an attempt to corroborate the redshift distances of Sc I galaxies, Halton Arp and David Block used the Tully–Fisher relationship (described earlier in this chapter) to calculate the distances to a representative sample of Sc and Sc I galaxies. In the case of the Sc galaxies, both distance determination methods were in good agreement. The Sc I galaxies, however, presented a major anomaly. The Tully–

Figure 6.14. If the Sc I galaxy NGC 309 is at its calculated redshift distance, it is one of the largest galaxies in the Universe. It would completely dwarf M81 (inset), which has traditionally been considered to be a large galaxy. (Photographs reproduced courtesy of Photolabs, Royal Observatory, Edinburgh.)

Fisher distances were smaller than the redshift distances by up to 100 million light years! As a final piece of evidence, Arp and Block calculated the rate at which supernovae should be observed in NGC 309, if it is indeed at its redshift distance, and hence calculated its gargantuan size. Based upon supernova rates in local galaxies, NGC 309 should present one of these celestial catastrophes every three years; but this is simply not observed, and the mystery of the Sc I galaxies continues.

Perhaps the explanation is somehow tied up with a documented but largely ignored observation dating from 1911. W.W. Campbell noticed that young, high-mass stars appear to have higher redshifts than expected. This is known as the K–Trumpler effect, and has been confirmed by subsequent observations with more sophisticated equipment in recent decades. An explanation, however, has not been forthcoming.

Since the discovery of quasars (which will be discussed in more detail in the following chapter) their nature has been debated. As far as their part in the battle to prove or disprove the Hubble expansion is concerned, everything rests on their redshift. Halton Arp conducted a survey of many peculiar galaxies, and discovered that a number of them appear to be connected to objects with very different redshifts. Arp's image of NGC 4319 and Markarian 205 (see Figure 6.15) may be the most important cosmological image of the century – outranking even the celebrated Hubble Deep Field! The image shows that the disturbed spiral galaxy NGC 4319 is apparently connected to the quasar-like object Markarian 205 by a luminous bridge of matter. The conflict occurs because a spectrum taken by Daniel Weedman showed that the recessional velocity of the quasar, based upon its redshift, was 21,000 km/s, as expected for a quasar. The galaxy, however, possess a recessional velocity of only 1,700 km/s. Again this is as expected, but how can the two then be joined by a bridge of material? Currently, there is no solution to this problem. It hangs like the sword of Damocles over the Hubble law. Stranger still is that no-one appears to be following up the investigation. Surely the reported bridge is a prime candidate for imaging with the superior optics of the Hubble Space Telescope.

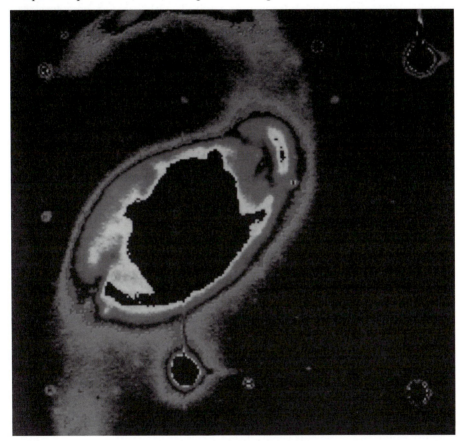

Figure 6.15. The quasar-like object Markarian 205 lies just below the disrupted spiral galaxy NGC 4319 in this image obtained by the addition of CCD frames taken by the Kitt Peak National Observatory's 4-m telescope. Note the apparently straight, luminous connection between the two galaxies. If verified, this feature would rock the foundations of the Big Bang theory, because the two objects possess radically different redshifts and hence should exist at dramatically different distances, precluding the possibility of their being joined. (Photograph reproduced courtesy of Halton Arp, from the cover of his book *Quasars, Redshifts and Controversies*, Interstellar Media, Berkeley, 1987.)

There are other reported links between galaxies and quasars. Some astronomers claim that quasars appear to be found around galaxies more often than predicted by statistics. The debate continues, but is as yet inconclusive. Some feel that gravitational lensing by stars in the outer reaches of galaxies may account for a brightening of distant, faint quasars. Sceptics opine that there is an insufficient number of faint quasars! Statistics can also be used to show that, if they are assumed to be chance alignments, far too many pairs of quasars with different redshifts are observed, based upon the total number of quasars known to exist. Some interpret this to imply that the quasar pairs must be physically associated – hence at the same distance – and that their different redshifts cannot be used as good indicators of that distance.

There is also evidence that perhaps the redshifts of galaxies are not smoothly and continuously distributed, as is indicated by the Hubble law. Data championed by William Tifft, of the University of Arizona, indicates that redshifts can exist only at certain values. This behaviour is very similar to the energy conditions which have to be satisfied by electrons in atomic orbits, and the ideas have become known as the quantised redshift hypothesis. This should not to be confused with the phenomenon of periodic redshift which is shown in data taken from pencil beam surveys. In these, a tiny area of sky is studied to great depth, so that the distribution of galaxies can be seen from their spread of redshifts. It has been found that galaxies fall into redshift groups separated by approximately 400 million light years of virtually empty space! This has been ascribed to the three-dimensional void and supercluster structure of the Universe at large. The periodic nature of the redshift corresponds to the survey piercing the galaxy-filled front and back walls of the voids.

There is still much which needs to be understood before astronomers can say that the Hubble flow has been proved beyond doubt. Although the Big Bang is still the theory of choice, it is important that the serious student of cosmology be aware that other theories exist.

6.13   OTHER COSMOLOGICAL THEORIES

To end this disquisition on scepticism of the currently accepted model, we shall briefly discuss two competing cosmological theories. The first of these – chronometric cosmology – was proposed by I.E. Segal, of the Massachusetts Institute of Technology. It posits a non-expanding Universe, and predicts a quadratic equation for redshift versus distance similar to the 1925 equation which K. Lundmark suggested could explain the apparent velocity, v, displayed by galaxies at a distance, d:

$$v = A + Hd + Bd^2 \tag{6.13}$$

where A and B are constants and H is the Hubble constant. Segal and a collaborator, J.F. Nicoll, showed that the change in magnitude with redshift displayed by the IRAS galaxies (see next chapter) was better fitted with a Lundmark relationship (equation (6.13)) than the Hubble law (equation (1.4)).

It is the assertion of chronometric cosmology that the Universe is curved. A fundamental constant represents the curvation factor in the same way that the speed of light delineates the relativistic regime and the Planck constant defines the quantum realm. This new constant is a length of 160 million parsecs $\pm 25\%$. Light is redshifted because it is travelling through a curved spacetime continuum; the further away the galaxy, the longer distance the light has had to travel, and so the more redshifted it will be. Hence, the redshift can still be used to measure distance, but by using equation (6.13) instead of equation (1.4).

Plasma cosmology is even more removed from the Big Bang theory. The plasma enthusiasts postulate that instead of gravity being the dominant force in the Universe, it is the electromagnetic force which is dominant. A plasma is a gas which is usually in a completely ionised state at high temperature. It can also be a partially ionised gas at a much lower temperature. A good analogy to a plasma is that of a metal at room temperature. In a metal, conducting electrons are free to wander around the fixed lattice of atomic nuclei. In a plasma, the situation with the conducting electrons is the same, and the atomic nuclei are also free to move about. Thus, due to the mobility of its electrons, a plasma is an excellent conductor of electricity. When it interacts with a magnetic field, large quantities of these electrons spiral around the field lines, causing a current to flow. These are known as Birkeland currents, named after the Norwegian scientist Kristian Birkeland, whose extensive research into plasmas forms the basis of the plasma cosmological theory.

Birkeland currents flowing between Jupiter and its moon Io were discovered by the Voyager space probe in 1979. In 1985 the first galactic Birkeland currents were discovered by Farhad Yusef-Zadeh and his collaborators whilst using the Very Large Array radio telescope to observe near the centre of our Galaxy. The Birkeland current they discovered is approximately 120 light years long, but only 3 light years wide. It is composed of a plethora of narrow filaments, and is associated with a magnetic field 100 times the strength previously thought possible on such a large scale.

Using this as their starting point, plasma cosmologists have shown that two typical interacting galactic Birkeland currents can emit energy which is almost equivalent in power to the radio galaxies (discussed in the next chapter). This may be significant, because the jets in radio galaxies are luminous due to synchrotron radiation which is caused by electrons spiralling around magnetic field lines: that is, synchrotron radiation is emitted every time a Birkeland current flows. If computer models are used to simulate two plasma clouds interacting with one another, double radio lobes can be produced which look remarkably like the radio lobes of active galaxies.

The cosmic microwave background also has a plasma explanation as computed by William Peter and Eric Lerner. They examined the behaviour of radiation released by plasma clouds in random directions throughout the Universe. Their results show that, under the right conditions – which may or may not be possible in the Universe at large – the radiation would possess a black-body spectrum appearing to emanate from a thermal source at just below 3 K. This is a good description of the cosmic microwave background.

Any plasma must have properties which differ from region to region. These regions are separated by transition zones. Some of the bolder plasma cosmologists are prepared to think that these regions can extend to larger and larger scales; and perhaps they are even the reason why the Universe is honeycombed with superclusters around giant voids.

In the plasma view of the Universe, redshifts are a result of the Wolf effect, which occurs in plasma clouds and 'drags' the frequency of emitted light down to a much lower level than would be natural. Thus, galaxies will exhibit redshifts, but these will not be indicators of distance; rather, they will provide clues about the plasma reactions taking place in the galaxy under observation. Wolf and his colleagues have already managed to simulate oxygen lines with a redshift of 0.07, produced by a stationary source! The approach looks very promising, and in future it may well compete strongly with cosmological redshift theories.

7

Active and IRAS galaxies

This book has so far referred to normal galaxies. Chapter 1 introduced the subset of galaxies which are known as active galaxies, and this chapter will expand considerably on the subject. As a broad average, about one galaxy in every ten is active. This activity originates in the core of the galaxy in question, and can be either directly observed in the central regions of the galaxy or inferred to take place there. Sometimes (as in the case of radio galaxies) the effects of the activity may only be visible much further away from the central regions of the galaxy. In most cases, however, the activity is caused by physical processes other than the radiation of energy from stars.

The study of active galaxies has been proceeding for the last 50 years, with intense activity over the last three decades since the discovery of quasars. Many research astronomers now investigate these exotic objects to determine their nature. Active galaxies are often bright landmarks in the far-distant Universe, and they provide some of the most efficient probes into those earlier epochs.

7.1 SEYFERT GALAXIES

Seyfert galaxies were first discovered by Carl Seyfert in 1943. He had been using the Mount Wilson Observatory redshift survey, and had distinguished six galaxies which each displayed some peculiarities in their spectra. What he had been looking for, particularly, were spectral emission lines. Normal galaxies do not display emission lines in their spectra, but the six which Seyfert had noticed showed broad emission lines. NGC 1068, NGC 3516, NGC 4051, NGC 4151 and NGC 7469 displayed a spiral morphology, but NGC 1275 was irregular. As more and more Seyfert galaxies have been discovered, this trend has continued. Most of them display spiral morphology. Observationally, a Seyfert galaxy appears like a normal spiral galaxy, but it contains a very bright nucleus which, if studied spectroscopically, displays broad emission lines.

As spectroscopic technology improved, more and more Seyfert galaxies were recognised. With increasingly detailed spectra being obtained for these objects, more subtle differences between them became obvious. The spectral lines, for example, were not uniformly broad. In 1974 two astronomers, Khachikian and Weedman, proposed a slight modification to the Seyfert classification scheme. These active galaxies would now be classified according to the width of their emission lines.

By this time, astronomers had noticed a difference in the types of spectral lines produced by celestial objects, which they categorised as permitted lines and forbidden lines (representative of the ease with which they can be reproduced under laboratory conditions). The forbidden lines are those transitions which are evident only in the highly rarefied environment of space. For example: in 1864 William Huggins discovered lines in the green part of the optical spectrum of an emission nebula, which led him to think that he had discovered a new element, which he named 'nebulium'. In 1927, Ira S. Bowen showed that these lines are actually those of doubly ionised oxygen [OIII] which can be produced only under conditions of extreme rarefaction.

In their studies of Seyfert galaxies, Khachikian and Weedman noticed that some examples of these active galaxies displayed differences between the widths of the permitted lines and the forbidden lines. Seyfert Type 1, as they called them, had permitted line widths which indicated Doppler velocities of 1,000–10,000 km/s; that is, the clouds of gas producing these emission lines were swirling around in the nucleus at these large velocities. The forbidden lines showed velocities of approximately 1,000 km/s. In the Seyfert Type 2 galaxies, permitted lines and forbidden lines displayed widths which indicated velocities up to 1,000 km/s (see Figure 7.1). It can be inferred that perhaps the broad lines – which indicate velocities of more than 1,000 km/s – and the narrow lines – which indicate velocities of less than 1,000 km/s – are generated in different parts of the active galactic nucleus (AGN). This initially allowed astronomers to talk confidently about broad-line regions (BLRs) and narrow-line regions (NLRs) without knowing anything about the physical characteristics of such places! However, astronomers have now developed an excellent working model for the central engine of an active galaxy which provides a very satisfying explanation of the cause of the distinct regions (discussed later in this chapter).

In recent years, technology has again necessitated the modification of the Seyfert classification scheme. Detailed study has shown that some of the Type 2 Seyfert galaxies have their narrow lines superimposed on much fainter, broader bases – almost as though the BLR is there, but not easy to see. Three new sub-classes are currently in use, and are based upon the appearance of the hydrogen lines in the spectrum. A Seyfert Type 1.5 is a narrow-line Seyfert which has broad bases to its hydrogen lines. Seyfert Type 1.8 has a distinctly broad base to its Hα emission line, but a fainter broad base to its Hβ line. Carrying this distinction one step further, a Seyfert Type 1.9 has a broad Hα base, but no Hβ base. So, instead of the rigid difference between the two types of Seyfert galaxy, a gradual blurring or merging of the types has become obvious with increasingly subtle observational techniques. This is perhaps our first clue that all types of Seyfert galaxies are intrinsically the same, and that the differences are caused by the way in which we perceive them. This idea will be pursued later in the chapter when we examine unification schemes for active galaxies.

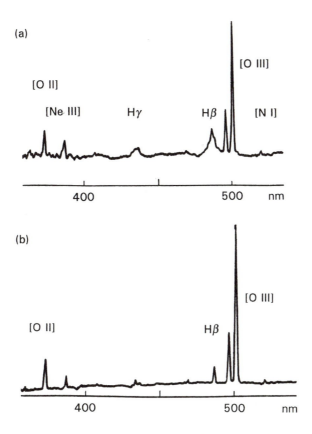

Figure 7.1. Optical spectra of (a) Seyfert Type 1 galaxy NGC 3227 and (b) Seyfert Type 2 galaxy Markarian 1157. Note the difference in the widths of the Hβ emission lines. (Adapted from Osterbrock, D., *Astrophysics of Gaseous Nebulae and Active Galaxies*, University Science Books, 1989.)

7.2   RADIO GALAXIES

The black-body curves possessed by thermal sources, such as stars, always tail off to include some radio emission. On top of this, if the other stars behave anything like our own Sun, surface phenomena such as flares and prominences will also produce non-thermal radio emission. Even so, the radio radiation emitted by a normal galaxy is rather low. Certain other objects in the Universe are copious emitters at radio wavelengths (more usually called frequencies when referring to this low-energy end of the electromagnetic spectrum). These are the expanding remnants of supernovae and certain types of active galaxy.

Radio galaxies were first discovered by the radio engineer and amateur astronomer Grote Reber. He had read the work of Karl Jansky, who was the first man to detect radio waves from space. Although Jansky, who was working under the auspices of Bell Telephone Laboratories, had made this fascinating discovery, his superiors had asked him to move onto other work, and his pioneering radio astronomy was abandoned. Reber reasoned that if the emission was black-body in nature, it should be more intense at higher frequencies. He therefore built a 10-m parabolic dish, which he operated from his back garden. At first he tried to detect radio at 3.3 GHz, and later at 910 MHz; both attempts were unsuccessful, and he instantly realised that whatever the source of radio from space, it was not a thermal, black-body source. In October 1938, he finally detected radio emissions from space at a frequency of 160 MHz.

Reber spent the next six years mapping the sky at this frequency, and he discovered that at 480 MHz he could clearly distinguish the band of the Milky Way. Within this, the centre of the Galaxy was very clearly defined as a strong source. He also acquired two other bright signals, neither of which corresponded to a bright optical source. These objects are now called Cygnus A and Cassiopeia A (after the constellations in which they are located).

More detailed work continued after the Second World War, and was rendered easier because the effort of developing radar systems had significantly improved the methods of radio detection. It was during this time that work on the identification of Cygnus A was proceeding apace. Several other bright radio sources had been discovered: most notably, Taurus A, Virgo A and Centaurus A. In the early 1950s, Taurus A was shown to be associated with the supernova remnant, the Crab nebula; Centaurus A was identified with the bright but dusty elliptical galaxy, NGC 5128 (see Figure 7.2); and Virgo A was shown to be associated with the elliptical galaxy M87. The decisive evidence for the identification of Cygnus A came when F. Graham Smith, of Cambridge, reduced the errors in the position of Cygnus A to 1 arcminute, which allowed Walter Baade and Rudolph Minkowski to use the 200-inch reflector at Mount Palomar to take a deep optical image of the field. They discovered a faint, sixteenth magnitude elliptical galaxy coincident with the point of radio emission. A clear picture was emerging: if these strong radio sources were not from supernova remnants in our own Galaxy, they were from distant galaxies, usually of the elliptical kind. Study of the radio spectrum of one of these galaxies indicates that the emission is produced by the magnetobremsstrahlung or synchrotron process (discussed in Chapter 2). The faster the electrons are moving, the more energy they can emit as they spiral around the magnetic field lines. The amount of radiation emitted is termed the flux density, S_f, and is linked to the frequency, f, by the following simple proportionality:

$$S_f \propto f^a \qquad\qquad\qquad (7.1)$$

where a is the spectral index, which varies between –0.5 and –3 for radio galaxies. The more negative the number, the 'steeper' the spectrum; the steeper the spectrum, the rarer the higher-energy photons of radio become, and so the radio galaxy is less powerful than a counterpart which has a 'flatter' spectrum (see Figure 7.3).

As technology has improved, so have the classifications and subclassifications which can be applied to radio galaxies. In general, however, the first and most obvious distinction is whether or not the radio emission is lobe-dominated or core-

Figure 7.2. The bright radio source Centaurus A is now known to be associated with this optical galaxy, NGC 5128. The galaxy possesses a dense dust lane that runs across its centre. (Photograph reproduced courtesy of AURA/NOAO/NSF.)

dominated. In the case of a lobe-dominated radio source, the emission derives from two gigantic regions which lie symmetrically on each side of the host galaxy. These are sometimes 'connected' to the central galaxy by jets of radio emission, and it is therefore obvious that the activity driving the phenomenon originates in the galaxy's nucleus. These lobes are usually extended by many arcminutes on each side of the central galaxy. When the distance to such a galaxy is taken into account, the actual linear dimensions of the lobes can be calculated. They are invariably enormous – typically extending for hundreds of thousands of parsecs; and in some cases they are millions of parsecs in length. These radio lobes are the largest single objects in the Universe, and are so gigantic that they often span the entire diameter of a galaxy cluster.

Lobe-dominated radio galaxies can be sub-classified according to the flux density at 178 MHz. This distinction was first shown by the Cambridge astronomers Barnard Fanaroff and Julia Riley. Radio galaxies are now grouped according to their Fanaroff–Riley (FR) class. FRI sources are the lower luminosity radio galaxies, their flux density at 178 MHz being typically below 5×10^{25} W/Hz. Their lobes appear to be very extended, but quite faint at their outer extremities, with the steepest spectra being displayed in these faint regions. They are connected to the central, optical galaxy by smoothly continuous jets.

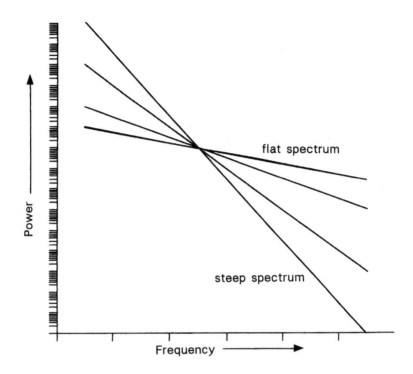

Figure 7.3. Flat and steep power spectra are designated according to the relative amounts of high-energy and low-energy radiation. Active galaxies which produce approximately equal amounts of high-energy and low-energy radiation are referred to as 'flat spectrum' objects.

FRII radio galaxies, on the other hand, are the most powerful sources of radiation. The steepest part of their lobe spectra is found in their inner regions, whilst the extremities are often edge-brightened. The jets are usually less comparable in brightness to the lobes than they are in FRI sources, and hence appear fainter than the lobes. In reality, both the jets and the lobes are significantly more luminous than their FRI counterparts. The jets are often not symmetrical in FRIIs; instead, one side will usually be far brighter than the other, or the source will be totally one-sided, with the jet feeding one lobe being totally invisible.

Core-dominated radio galaxies do not display these huge lobes of radio emission, but some of them do possess single-sided jet structures which can extend over distances of a few kiloparsecs. The radio spectra of these sources are usually almost flat, and extend into the millimetre/submillimetre region of the electromagnetic spectrum, to frequencies in excess of 300 GHz.

Figure 1.5. Supernova 1987A, in the Large Magellanic Cloud. Supernovae have processed the chemical constituents of the Universe throughout its history. They have converted a small proportion of the hydrogen and helium into the heavy elements. Smaller stars that have not become supernovae have also contributed to the Universe's complement of carbon. The double loop structure observed around this supernova is quite mysterious, and has yet to be satisfactorily explained. (Photograph reproduced courtesy of Hubble Heritage Team (AURA/STScI/NASA).)

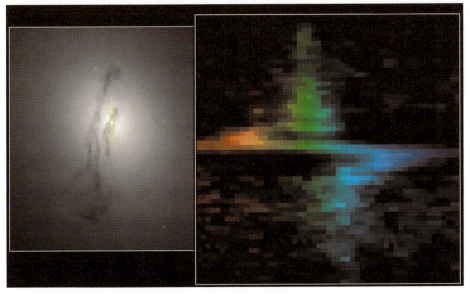

Figure 7.11. The spectrum (right) of the central region of the active galaxy M84 (left) in the Virgo cluster, showing one spectral line. The distortion is caused by the Doppler effect on the line produced by the rotation of the gas in a disc surrounding a central black hole. The central wavelength of the line has been colour-coded green to accentuate the redshift and blueshift. (Image reproduced courtesy Gary Bower, Richard Green (NOAO), the STIS Instrument Definition Team, and NASA.)

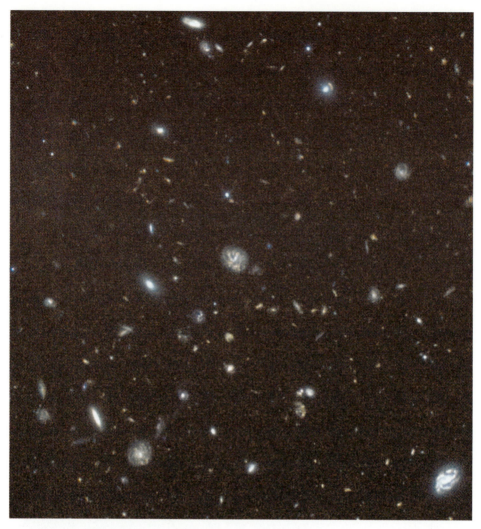

Figure 9.6. The Hubble Deep Field South is a composite colour image obtained with the Wide Field and Planetary Camera on the Hubble Space Telescope. It is a tiny region of sky in the Hubble South Continuous Viewing Zone, and lies in the constellation Tucana. The image is very similar in appearance to the Hubble Deep Field North, and so provides evidence that these are representative portions of the entire Universe and that the cosmological principle holds true. (Photograph reproduced courtesy of R. Williams (STScI), the HDF-S Team, and NASA.)

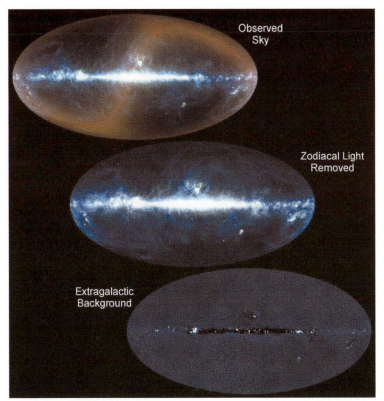

Figure 9.8. The infrared background – the all-sky map constructed from results obtained by the DIRBE instrument on board the Cosmic Background Explorer (COBE). Removal of the zodiacal and Galactic emission leaves a residue of infrared radiation which has been interpreted as emanating from the high-redshift population of protogalaxies. (Photograph reproduced courtesy of Michael Hauser (Space Telescope Science Institute), the COBE/DIRBE Science Team and NASA.)

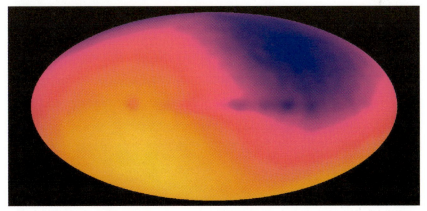

Figure 10.1. The COBE dipole anisotropy. This all-sky microwave map is plotted in galactic co-ordinates, with the plane of the Milky Way running horizontally across the middle. It shows variations in the effective temperature of the cosmic microwave background radiation. The motion of our frame of reference imparts a dipolar Doppler shift upon the cosmic microwave background radiation. The dipole is present at the level of 1 part in 1,000. (Photograph reproduced courtesy of COBE Science Working Group, NASA, GSFC, NSSDC.)

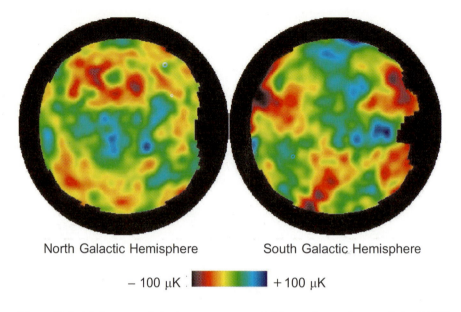

North Galactic Hemisphere   South Galactic Hemisphere

– 100 μK   + 100 μK

Figure 10.4. These maps of ripples were constructed from observations made by COBE over a period of four years. They are projected onto Galactic polar maps – one for the north Galactic pole, and the other for the south Galactic pole. The ragged edges are produced by the masking of unreliable data from the plane of the Milky Way. These maps represent the distribution of matter at the point of decoupling. (Photograph reproduced courtesy of COBE Science Working Group, NASA, GSFC, NSSDC.)

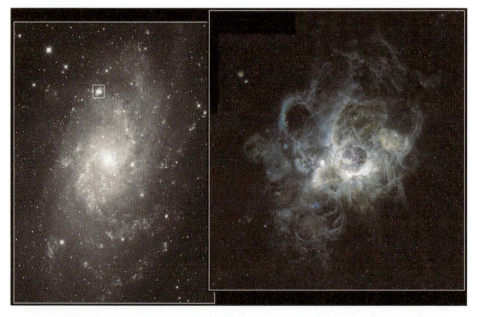

Figure 10.8. NGC 604 is an enormous emission nebula in an external galaxy. Nebulae as large as this – and even larger – must have been a feature of the Universe when galaxies were first forming. Such high-mass stars reionise the interstellar medium. (Photograph reproduced courtesy of Hui Yang (University of Illinois) and NASA.)

7.3   QUASARS

The great interest which followed the discovery of the radio galaxies encouraged workers in the newly burgeoning field of radio astronomy to attempt the first sky survey. This was collated at Cambridge by Martin Ryle, F. Graham Smith and Bruce Elsmore in 1950, and resulted in the *First Cambridge Catalogue*, containing 50 northern hemisphere radio sources. Each object was referenced as IC, followed by the source's catalogue number. Five years later, the *Second Cambridge Catalogue* was published, with almost 2,000 sources. As radio astronomy continued to become more and more sophisticated, it was realised that much of the 2C catalogue had resulted from overzealous plotting of sources; and so, in 1962 the third and definitive catalogue was issued. This contained only a quarter of the sources included in the 2C catalogue.

As already mentioned in this chapter, many of these sources were discovered to be extended objects which fell into two categories. Each of them was either a supernova remnant or a radio galaxy. A certain number of objects were not extended, and appeared to be point-sources of radio emission. It is slightly misleading to term these objects 'point-sources', because the beam widths of those early telescopes were so large that any object smaller than several arcminutes in extent was considered to be a point source. Nevertheless, these point-source objects were intriguing; they were relatively strong radio sources, and were ripe for further investigation. The strength of the emission and the compact nature of the sources led many to believe that they were nearby 'radio stars' within our own Galaxy; but the distribution of sources throughout the sky precluded this argument. Instead of being confined to the Milky Way, the sources were dotted (more or less) isotropically across the sky. This indicated to some that the sources must be extragalactic in origin. Others continued to argue that, statistically, the isotropy was not that significant.

The only way to solve this puzzle was to determine exact positions for the point-sources, and then to determine if any optical object could be found and studied. With the available data, the positions were so inaccurate that when correlated with optical plates, any one of 100 visible objects could have been the optical counterpart. In an effort to refine the positional accuracy of these point-sources, the new technique of radio interferometry was employed. Jodrell Bank astronomers showed that 3C48 was indeed a point-source by demonstrating that the radio emission was confined to an area of the sky which was less than one arcsecond in diameter. Astronomers at the California Institute of Technology also began investigating this object in order to refine its position. Eventually, Allan Sandage – using a photographic plate taken whilst employing the 200-inch telescope on Mount Palomar – identified it with a sixteenth magnitude blue stellar object which was surrounded by some faint nebulosity. Further investigation began immediately. The object was studied photometrically, and its spectral colours were found to be very peculiar – quite unlike anything previously observed in a star.

The next step was to study the star spectroscopically. The mystery deepened. The spectrum contained broad emission lines, a feature not previously seen in stars, and attempts to identify the spectral lines with chemical elements met with failure. It

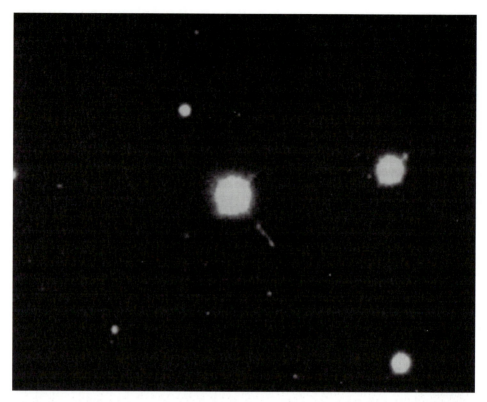

Figure 7.4. 3C273, in Virgo, is the brightest quasar in the sky, and is a strong radio source. This Kitt Peak National Observatory 4-m telescope photograph shows the visible part of this thirteenth magnitude object, including the prominent jet. We see only the point-like active nucleus, which dramatically outshines the stars in the underlying 'host' galaxy. This quasar may radiate 100 times more light than the brightest ordinary galaxy, and its jet may measure 150,000 light years in length. (Photograph reproduced courtesy of NOAO/Kitt Peak National Observatory.)

seemed as if none of them matched with anything previously known. So just what were these objects which had the appearance of stars, but displayed none of the characteristics of any known celestial object?

Investigations of 3C48 led many to believe that what had been discovered were radio stars. The story continues, however, with the work of Cyril Hazard, the English astronomer who led the first successful discovery of the true nature of these objects. Hazard was working in Sydney, Australia, using a radio interferometer. He calculated that, of the various compact radio sources in the Cambridge catalogue, 3C273 (see Figure 7.4) would be occulted by the Moon. This would occur not just once but three times during 1962. If he could observe this occultation, he could pinpoint the position of 3C273 by tracing precisely when the radio emission was blocked by the Moon.

Figure 7.5. z = 5 quasar discovered by the SLOAN Digital Sky Survey. The light from this quasar has been travelling through space for approximately 14 billion years, having left the quasar when the Universe was about 1 billion years old. This is the most distant quasar yet discovered. This image was obtained with the Japanese 8-m telescope, Subaru. (Photograph reproduced courtesy of National Astronomical Observatory of Japan.)

The newly established position allowed Hazard to use a Mount Palomar optical plate, brought to Australia by Rudolph Minkowski, to identify 3C273 as a thirteenth magnitude object which resembled a star. This information was communicated to Mount Palomar where, in December of 1962, Maartin Schmidt obtained an optical spectrum of the source. It displayed the same broad emission lines as 3C48, and with similarly intractable patterns. Schmidt was persistent in his efforts to understand this spectrum, and he eventually recognised the Balmer lines of hydrogen. The only remaining puzzle was that they were nowhere near where they were expected to be in the spectrum. They had been severely redshifted. Taking this extraordinary result at face value, Schmidt calculated a distance to the object using the Hubble law with a constant of 50km/s/Mpc. It was an extraordinarily large 948 Mpc. Calculating the luminosity of 3C273, it was found that it is about 50 times more luminous than even the brightest galaxies (five million million times the luminosity of our Sun)! Thus, quasars were not stellar objects at all. Instead they were the brightest form of active galaxy known, and existed at extreme distances. Because of their appearance they became known as quasi-stellar radio sources – quasars. Today quasars are the most distant objects that can be seen in the Universe (see Figure 7.5). Currently, the most distant is known as Quasar 1. Its redshift is a staggering 5, placing it almost at the very edge of the observable Universe. When the light that has been observed from this quasar began its journey through space, the Universe was 90% younger than it is today (see Figure 7.6).

The broad emission lines in quasars suggest Doppler velocities of up to 10,000 km/s. In many cases, the spectra are very similar to those of Seyfert Type 1 galaxies. Although quasars were originally discovered due to their strong radio emission, subsequent work at optical wavelengths has shown that the vast majority of them – about 90% – are radio quiet. These were initially term QSOs for quasi-stellar objects, although this subclassification seems to have fallen from common usage, and both radio-loud and radio-quiet varieties are now referred to as quasars. If any

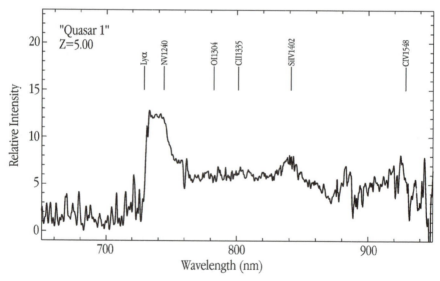

Spectrum of Quasar at Z=5.00 (VLT UT1 + FORS1)

Figure 7.6. The spectrum of the $z = 5$ SLOAN quasar – now called Quasar 1 – was obtained with the Focal Reducer Spectrograph (FORS) on the first unit telescope of the VLT. The exposure lasted for an hour. This is an astounding feat, and is a tribute to the quality of both the FORS and the VLT. (Photograph reproduced courtesy of ESO.)

qualification is needed, it will usually be explicitly stated that the quasar has a radio-loud component in its spectrum.

The optical properties of quasars are, in general terms, so similar to those of Seyfert Type 1 galaxies that the distinction between the two has become extremely blurred. For instance, the lowest-luminosity quasars are less powerful than the highest-luminosity Seyfert Type 1 galaxies. A rather arbitrary, but workable, scheme has been the segregation of these objects by their redshifts. Those with a redshift of less than 0.1 are classified as Seyfert Type 1, and those with a redshift between 0.1 and 5 are termed quasars, regardless of intrinsic luminosity.

Just as the only distinguishing feature of a Seyfert galaxy is the overly luminous central nucleus, it has always been assumed that quasars are the super-bright nuclei of otherwise normal galaxies. This theory was recently confirmed by the Hubble Space Telescope in a set of observations conducted by John Bahcall (Institute for Advanced Study, Princeton) and Mike Disney (University of Wales). The results show that quasars can reside in apparently undisturbed, normal galaxies, which can be either spiral or elliptical in shape. Approximately half of the quasars studied, however, were found to exist in merging systems (see Figure 7.7). This has important

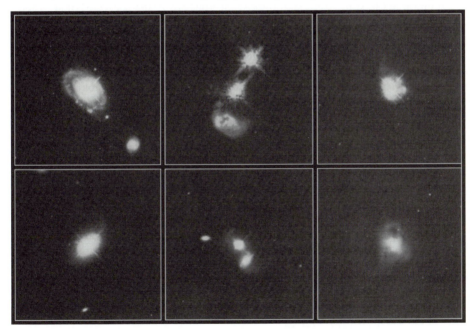

Figure 7.7. Six quasars and their host galaxies can be seen in these images obtained with the Hubble Space Telescope. Top left is quasar PG 0052 + 251, which can be seen to be in a normal spiral galaxy. Below this is quasar PHL 909, which resides in a normal elliptical galaxy. The other four quasars are all in galaxy mergers. These are: centre. IRAS 04595-2958; bottom centre, PG 1012 + 008; top right, quasar 0316-346; bottom right, IRAS 13218 + 0552 (Image reproduced courtesy of John Bahcall, Institute for Advanced Study, Princeton, Mike Disney, University of Wales, and NASA/STScI.)

implications when astronomers consider how active galaxies relate to one another and, in particular, how an active galaxy generates its tremendous energy output. This will be discussed in the section on unification schemes.

7.4　BLAZARS

The final category in our survey of active galaxies is the group which has become known generically as blazars. In 1941, an object in the constellation Lacerta was classified as a variable star by H. van Schewick. In accordance with variable-star nomenclature, it was called BL Lacertae. The 'star' returned to the centre of the astronomical community's attention in 1968 when it was discovered – by a survey made at the University of Illinois' Vermillion River Observatory – to be a strong radio source. Further investigation revealed that it had very little in common with a star. Its spectrum did not display the expected stellar absorption lines, or the more unusual emission lines. A study of the spectral energy distribution, (the power emitted at the different wavelengths) revealed that these radio sources have very little in common with stars, but share most of the properties of the (then recently discovered) quasars. Thus, in an act of faith, astronomers classified BL Lac objects as a new type of active galaxy. As techniques have improved, so emission lines have been found in a small number of BL Lac objects, which has proved that these objects are indeed extragalactic and have maximum luminosities which overlap with the minimum for quasars.

BL Lac objects are highly variable in nature, fading from brilliance before returning to prominence, on irregular time-scales that can sometimes be measured in days. The emission is also highly polarised, which indicates that the radiation is being produced by high-velocity electrons spiralling around magnetic field lines to produce synchrotron radiation.

A subset of quasars, known as the OVVs (optically violently variables) was noted to have certain properties in common with the BL Lac objects: both display highly polarised radiation and are very variable. Similarities in the spectral energy distribution has also led astronomers to concatenate the two groups and amalgamate the names BL Lac and quasar into blazar. The OVV component of the blazars extends the high end of the luminosity range into a broad overlap with the more general quasars.

7.5　THE CENTRAL ENGINE

The question of just what could possibly provide so much energy for the active galaxies to radiate into space was a natural one which sprang to the minds of the astronomers who studied these fascinating objects. The early 1960s was a time of great speculation about these objects' power-houses. As a broad generalisation, these objects existed at earlier epochs; that is, at greater redshifts. This single fact led many to speculate that the cause of this excess radiation was a massive bout of star formation. Galaxies in the present-day Universe have nuclei populated with old stars; formation has long since ceased and is now confined to some irregular and peculiar galaxies and to the spiral arms of certain galaxies. Although, as we shall see in our discussion of starburst galaxies, this theory cannot be completely dismissed, it was superseded by another 'astronomical child of the '60s'.

In the same way that active galaxies and quasars suddenly became elevated in importance because of the advances in technology during the 1960s, so also did black holes. Prior to this decade, they were confined to the scientific curiosity box. General relativity seemed to be the only way in which to study these theoretical conundrums. In 1969, however, Donald Lynden Bell, from Cambridge, proposed that a quasar could be powered very efficiently if it possessed a black hole at its core. Any object, such as a star, which strayed too close would be ripped to shreds by the tidal forces of the black hole. The energy which can be liberated by an object falling into a black hole is enormous. It is the most efficient energy production method in the Universe, converting 42% of the object's mass into energy.

The black holes found in active galaxies have been termed supermassive black holes, to distinguish them from their smaller cousins which are thought to be produced in supernova explosions. Supermassive black holes can contain more than 100 million times the mass of the Sun, and be no larger than 2 AU in diameter. In other words, the central engines of the brightest objects in the Universe may be no bigger than the diameter of the Earth's orbit! This size estimate is supported by observations which measure the variability of the radiation output from the active galactic nucleus. The radiation observed to be coming from an active galaxy obviously cannot be emanating directly from the black hole. Instead it is thought to come from a disc of gaseous material which is spiralling down into the black hole. Some active galaxies have been measured to vary their radiation output on a time-scale of a day or so. In order to compensate for any areas of the accretion disc which are decreasing in brightness as other areas are rising, it is assumed that a change in the output of the source corresponds to the brightening of the whole emitting region. This brightening must propagate across the disc at a velocity which cannot exceed the speed of light. So, taking these two conservative assumptions at face value and combining them with the rate of variability, the maximum size of the emitting region must be no greater than the distance that light would travel in a single day. This distance is about the same as the diameter of the Solar System, which is the expected size of the accretion disc around a supermassive black hole.

The accretion disc forms for essentially the same reason that water spirals down a plug-hole. It is the rotation of the Earth which causes bath water to spiral downwards. The mass of the black hole is so great that it creates an incredibly deep gravity well in the centre of the galaxy (the cosmic plug-hole). As material falls down the gravity well it is still moving in its orbit around the centre of the galaxy (which probably corresponds to the position of the black hole). As the gaseous material orbits the black hole it will interact with other clouds in similar but differently oriented orbits, and eventually will settle into an accretion disc because of the energy it has lost in the collisions. The gas in the accretion disc is heated to incandescence by the shearing forces and friction in the disc. This causes it to emit even more radiation

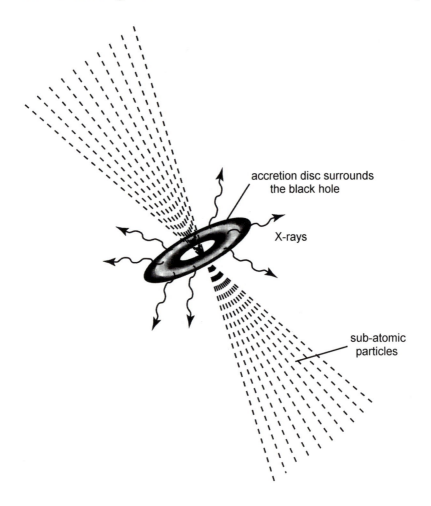

accretion disc surrounds
the black hole

X-rays

sub-atomic
particles

Figure 7.8. It is possible that the central engine of an active galaxy is an accreting black hole. As the black hole devours stars and gas clouds, intense amounts of radiation are emitted by the swirling gas.

in the form of X-rays, because the gas particles are heated to temperatures in excess of 10^5 K (see Figure 7.8).

Using this model, the broad-line region can then be explained quite naturally as the re-radiated emission from the swirling gas clouds which are yet to settle into the accretion disc. The details of the model are, however, far from simple. The gas which radiates from the narrow-line region exists much further out, at a distance of 10–100 parsecs from the black hole, where the radiation and kinetic energy are less intense.

Another line of investigation which builds astronomers' confidence that black holes are present in the core of active galaxies, is the motion of emitting gas in the near vicinity of the galaxy's centre. In the case of M87, a radio galaxy in the Virgo cluster, Holland Ford (Johns Hopkins University) used the Planetary Camera

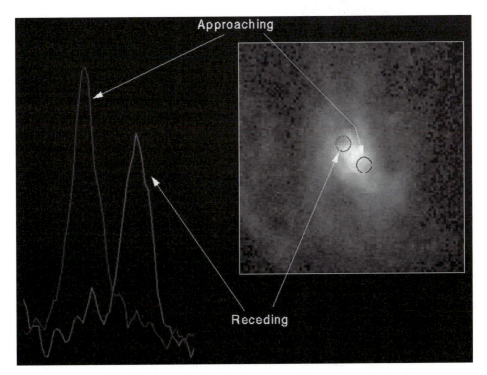

Figure 7.9. The image at right shows a conspicuous disc-like feature in the centre of the active galaxy M87. The spectrum – taken at two opposite points in the disc – clearly shows that the spectral lines have been Doppler-shifted. On one side of the disc the spectral line is blueshifted, and on the other it is redshifted. The value of the Doppler shift allows the velocity of the gas to be calculated and the mass of the central black hole to be deduced. (Image reproduced courtesy Holland Ford, Space Telescope Science Institute/Johns Hopkins University; Richard Harms, Applied Research Corp.; Zlatan Tsvetanov, Arthur Davidsen, and Gerard Kriss at Johns Hopkins University; Ralph Bohlin and George Hartig at Space Telescope Science Institute; Linda Dressel and Ajay K. Kochhar at Applied Research Corp. in Landover, Maryland; and Bruce Margon from the University of Washington in Seattle, and NASA.)

onboard the HST to capture an image of an ionised gas disc in the innermost few arcseconds of the galaxy. Then, Richard Harms (Applied Research Corporation) followed up with the HST's Faint Object Spectrograph to capture spectra of the gas on each side of the disc (see Figure 7.9). The Doppler effect of the gas clearly showed that on one side of the disc the gas was moving towards Earth, whilst on the other it was moving away. Typically, the speed of the gas was more than 450 km/s. It was in orbit around a central, massive object. When the mass of the gravitating body was calculated, it was found to be $\sim 3 \times 10^9$ M_{Sun}. At this mass, it can only be a black hole squeezed into the central confines of the galaxy.

Other active galaxies have since fallen under the gaze of the HST. NGC 7052 was found to have a dusty disc-like structure, 3,700 light years in diameter, with gas

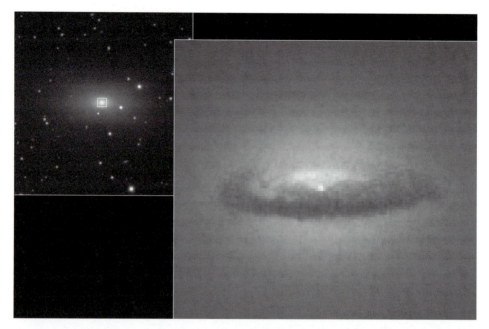

Figure 7.10. This enormous disc of dust is 3,700 light years in diameter, and encircles a 300 million solar-mass black hole in the centre of the elliptical galaxy NGC 7052. It may be the remnant of a cannibalised galaxy from NGC 7052's past. It is estimated that all the matter in this disc will have been consumed by the black hole within a few billion years. (Image reproduced courtesy Roeland P. van der Marel (STScI), Frank C. van den Bosch (University of Washington), and NASA.)

moving around the centre of the galaxy at 155 km/s. These figures again implied that a 3×10^9 M_{Sun} black hole was to be found in the centre of the galaxy. It would account for 0.05% of the galaxy's mass (see Figure 7.10).

Another galaxy, M84, was observed with the Space Telescope Imaging Spectrograph (STIS) shortly after its installation during the second HST servicing mission. The beautiful 'zig-zag' spectrum that was obtained showed gas moving with a velocity of 400 km/s within the central 26 light years of the galaxy (see Figure 7.11, colour section). Again, this implied the existence of a 3×10^9 M_{Sun} black hole, and a cannonical mass of a supermassive black hole was beginning to become apparent.

7.6　UNIFICATION SCHEMES

An obvious question to ask is: why are some active galaxies more powerful than others? The immediate answer would appear to be that if a supermassive black hole generates the energy output in each active galaxy, then the power output will be related to the amount of matter being accreted. In the case of the most powerful active galaxies, the quasars, theoretical calculations have shown that about two solar masses of material must be accreted every year.

An obvious way to deliver a large amount of matter into the centre of an active galaxy is for it to collide with another galaxy. Hence, the observation that half of the quasars are in merging systems appears to prove this assertion. Of the quasars that appear to be in undisturbed galaxies, the obvious answer is that two galaxies have indeed collided, but as one is so much smaller than the other, it has not disrupted the external appearance of the larger galaxy.

Dramatic evidence that the nearest powerful active galaxy is powered by a collision was obtained by ESA's Infrared Space Observatory (ISO) in 1998. Centaurus A is a peculiar radio galaxy that radiates as much energy as a quasar. In optical light, it is striped by an enormous dust lane which runs across the centre. The infrared camera on board ISO captured an image of this dust lane at a wavelength of 7 μm, revealing that it was once a spiral galaxy that is now passing directly through the centre of Centaurus A (see Figure 7.12). This titanic galactic collision is feeding a supermassive black hole at the centre of the merging galaxies, and is creating the active phenomena.

Other active galaxies – such as Seyferts – are not part of merger systems, and their

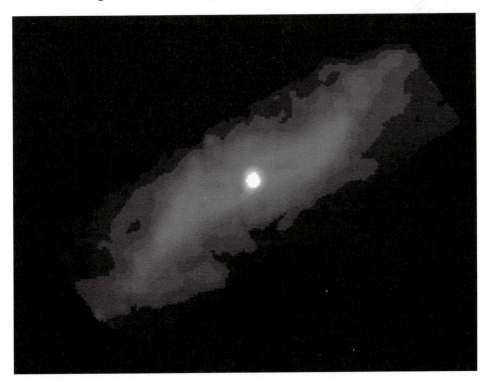

Figure 7.12. When observed with the European Space Agency's Infrared Space Observatory (ISO), the dust lane of Centaurus A is revealed as a flat, spiral galaxy that has collided with the giant elliptical galaxy. The host elliptical galaxy is not seen in this view. (Image reproduced courtesy ESA/ISO/ISOCAM, CEA-Saclay and I.F. Mirabel *et al.*)

power output is maintained by a supply of gas that 'leaks' into the heart of the galaxy. As yet, the exact mechanism of how the gas loses angular momentum and falls into the centre of the galaxy is unclear, and is the subject of intense research.

In our initial descriptions of the different types of active galaxy, it was noted that some display broad emission lines and others display only narrow emission lines. We next consider the question of what causes the active galaxy to be a broad-line galaxy or a narrow-line galaxy.

One of the most important observations of active galaxies was made in 1985, when Roberto Antonucci and Joe Miller observed the Seyfert Type 2 galaxy NGC 1068 in polarised light. Instead of narrow lines they saw broad lines; the galaxy did have a BLR, but it could not be seen in direct light. In Chapter 2 it was stated that one mechanism by which light can become polarised is when it is scattered off dust clouds. In some way, the BLR region of NGC 1068 is hidden from our direct line of sight, but small quantities of light from it are scattered in our direction by dust clouds.

The broad-line region must be surrounded by a torus of dusty material, about 1 parsec in diameter, which, from certain angles blocks our direct view of the central regions (see Figure 7.13). The edges of this torus must be optically thin enough, however, to allow light to scatter in our direction. The narrow-line region then exists beyond this torus. Corroborative evidence that Seyfert galaxies may all be intrinsically the same type of object has come from NGC 4151, which has been observed at different epochs to change its observational properties from that of a Seyfert Type 1 to a Type 1.8 and back again. Perhaps this is caused by precession of a non-uniform torus. When the torus partially blocked our view, the broad lines almost disappeared, and it was observed as a Type 1.8. It then returned to its broad-line status after the obscuration passed.

We can extend this idea of unification to the radio-quiet quasars because there is an overlap between the luminosities and properties of quasars and Seyfert Type 1 galaxies. As previously stated, one of the basic classification criteria is the artificial consideration of the object's redshift. Thus, perhaps these three types of active galaxy are indeed intrinsically the same type of object, and overzealous classification has blinded us to the fact. The only difference between these types of object is distinguished by our ability to see over the obscuring torus and into the BLR to the power of the radiation being emitted by the central engine.

Radio-loud varieties of active galaxy can also be organised into a unified sequence. The radio-loud quasars are our starting point in this scheme. One of the characteristics of radio-loud AGN is their association with jets of material. From the high polarisations measured, the jets must almost certainly be funnelled out of the AGN by magnetic fields. A quasar displays broad-line emission, and so our viewing angle must be well away from the torus. If the angle becomes steep enough, however, our view of the broad-line region will be obscured by the jet emission. This is the scenario proposed for blazars. Astronomers are literally blinded by the light, because they are observing along the jet's axis of symmetry. Emission lines are therefore incredibly difficult to detect.

Tilting our line of sight in the other direction will eventually lead us to lose sight

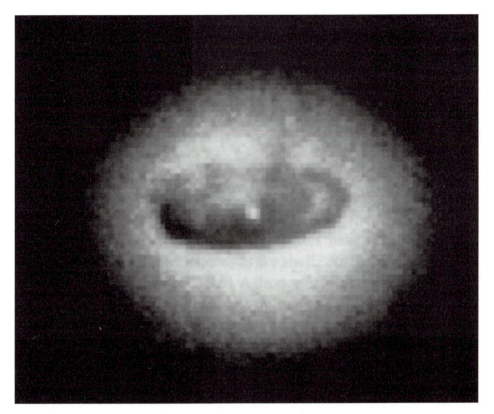

Figure 7.13. The core of the active galaxy NGC 4261. This image shows a torus of dust, 800 light years across, fuelling a supermassive black hole at the centre of the galaxy, which is located at a distance of at least 100 million light years in the direction of the constellation Virgo. It has been calculated that the object at the centre of the disc is 1.2 billion times the mass of our Sun, concentrated within a region of space not much larger than our Solar System. (Photograph reproduced courtesy of L. Ferrarese, Johns Hopkins University and NASA/STScI.)

of the broad-line region because it is obscured by the torus. In this case, the active galaxy will appear to be a radio galaxy (see Figure 7.14).

If we wish to extend the ideas of unification, we have to consider a way in which radio-loud and radio-quiet galaxies can be brought together. This is a much more difficult job, and many astronomers are of the opinion that it will remain a fundamental difference between active galaxies. A curious observational aspect is that the majority of radio-quiet active galaxies appear to be hosted by spiral galaxies. Conversely, the observational impression produced by studies of radio-loud galaxies is that they are embedded within elliptical galaxies which in some way are often disturbed.

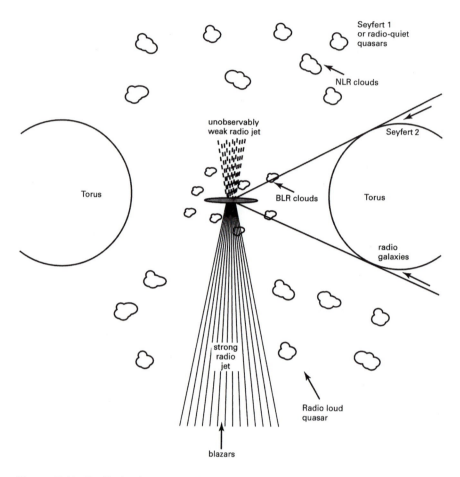

Figure 7.14. Radio-loud and radio-quiet unification – possible scheme for the unification of active galaxies. The upper and lower halves of the diagram are separate entities. The upper half shows radio-quiet unification, and the lower half shows radio-loud unification.

7.7   IRAS GALAXIES

Our discussion so far has been about the traditional active galaxies. Recently, astronomers have been wondering just how these fit in with the so-called 'normal' galaxies that we see around us in the present epoch. Investigations during the 1980s also revealed another class of galaxy – the IRAS galaxies, which have instigated a reappraisal of galaxies and the interrelations.

IRAS (Infrared Astronomical Satellite) was launched in 1983. It revolutionised infrared astronomy by being the first instrument to perform an (almost) all-sky survey (95%) at wavelengths of 12–100 μm. During the mission (which lasted for just under a year) IRAS detected approximately 250,000 infrared sources scattered throughout the Universe. In the years that followed, most of those sources were identified with objects, particularly stars, within our own Galaxy.

The beam size used by IRAS was two arcminutes, and anything which appeared smaller than this was classified as a point-source. Obviously, all of the stars detected came from the point-source catalogue. Of the remaining point-sources, a significant number were correlated with external galaxies. Further investigation showed that they were completely normal spiral galaxies which had been detected by IRAS. In some cases the infrared radiation is 50 times more luminous than the optical emission. This was instantly fascinating, because the wavelengths selected for IRAS to study were purposely chosen because individual stars emit relatively little radiation at this longward end of the waveband.

Stellar sources radiate black-body curves with characteristic temperatures of 3,000–50,000 K. Peak radiation between the wavelengths of 60–100 μm requires a characteristic black-body temperature of 25–100 K. Thus, the majority of the emission perceived in this range was presumed to be coming from dust within the detected galaxies which had been heated by starlight from young stars. A surprising result was that a small fraction of the spiral galaxies detected by IRAS were found to have luminosities which rivalled the quasars. Their spectra, however, were not at all similar to those of the active galaxies. Instead, they resembled the spectrum displayed by an HII region – an emission nebula which signifies the site of star formation. Other galaxies detected by IRAS were slightly more modest, and favoured luminosities which were comparable with Seyfert galaxies. Further studies have shown that the number of IRAS galaxies actually far outweighs the number of Seyfert galaxies detected in the Universe.

As is always the way in astronomy when a new class of celestial object is discovered, a number of different theories sprang up to explain the phenomenon. Some thought that the IRAS galaxies (as they became known) were quasars in which the active galactic nucleus was enshrouded in dust. Others felt that they were galaxies in which an incredibly active bout of star formation was taking place. These galaxies are known as starburst galaxies.

As previously stated, stars in the present-day Universe are only 'naturally' created in the spiral arms of spiral galaxies and in certain forms of irregular galaxy. In both cases, star formation takes place at a regular, steady pace. Stellar masses can vary between 0.05 M_{Sun} and 60 M_{Sun}. The smaller the mass, the greater the numbers. High-mass stars are incredibly bright, with luminosities in excess of 10,000 times that of the Sun.

In other situations, stars can be 'forced' to develop in galactic collisions. When two galaxies merge it is rather like the interlinking of fingers. Although stars do not collide with each other because the spaces between them are so great, the giant molecular clouds do merge. The increased density results in an abnormally large fraction of the merged clouds collapsing gravitationally to become stars. This is the starburst phenomenon, and although it is a common event in interacting galaxy systems, it is by no means confined to them. The galaxy M82 in Ursa Major is a good example of a starburst occurring in a relatively close-by galaxy (see Figure 7.15).

Figure 7.15. M82 is the starburst galaxy nearest to the Milky Way. Its irregular shape indicates that some measure of disturbance has taken place, leading to the burst of star formation taking place within. (Photograph reproduced courtesy of Photolabs, Royal Observatory, Edinburgh.)

Although the galaxy is disturbed in appearance, the exact cause of the starburst is, in this case, not certain. One possibility is that the galaxy has been tidally disrupted by the near-by spiral galaxy M81.

7.8   ARE IRAS GALAXIES ACTIVE GALAXIES?

The perplexing question – which requires an answer if we are to gain an insight into galaxy evolution – is: are IRAS galaxies active galaxies? Certainly, they are not normal, but when the word 'active' is used in this context, what is really meant is: are they powered by a black hole accreting mass from its surroundings? At present, the IRAS galaxies are thought to be starbursts. Although some have been observed to be interacting, a significant proportion of them are isolated field galaxies, and it is not yet understood why starbursts should be taking place in these non-interacting systems. Whatever the precise explanation for the reason behind the burst of star formation, the single most obvious fact is that the starburst is hidden from direct view by a surrounding cocoon of dust. The young starburst cluster at the centre of the galaxy must contain many hot young stars. These emit predominantly in the blue and ultraviolet region of the electromagnetic spectrum, and observations clearly indicate a lack of these wavelengths. The atomic gas spectra of these IRAS galaxies also supports the fact that the highest-energy photons are 'mopped up' before exciting the gas into highly ionised states. Instead of high-energy 'blue' emission, the galaxies display excessive amounts of 'red' which, in collaboration with the other phenomenon mentioned above, is entirely consistent with the ultraviolet photons being absorbed by dust which then re-emits the radiation at infrared wavelengths. A similar process actually takes place on the Earth, whereby the ground is heated all day by the high-energy photons of radiation from the Sun. At night, the ground re-radiates this energy as longer-wavelength photons of infrared.

So if IRAS galaxies do not have black-hole cores, what makes them so fundamentally different from active galaxies? In order to begin answering that question we shall examine our basic assumption: that active galaxies and IRAS galaxies are powered by different mechanisms. Astronomers originally became suspicious of the active galactic nuclei interpretation because, as previously mentioned, the spectral energy distribution and emission-line spectra looked so different. They resembled, more strongly, the spectra of star-forming HII regions within our own Galaxy.

A group of astronomers led by Roberto Terlevich, of the University of Cambridge, began to investigate a property of quasars which was seldom considered to be significant. In fact, a result was predicted which has cosmological significance, and shows how useful these active galaxies are for probing the early Universe. In the spectrum of many quasars are emission lines which are due to heavy elements. In Chapter 4 we described how the Big Bang caused the Universe to be filled with atomic matter, the predominant fraction of which was hydrogen and virtually all of the remains was helium. Tiny traces of lithium were also formed. All of the other chemical elements have been formed in the centres of stars and released back into space at the end of the stars' lives. These heavy elements – the metals, as they are called – are present in the spectrum of even the youngest quasars. The conclusion is inescapable: for the lines to be as strong as they, large numbers of stars must have lived and died in the quasar before the epoch at which we are observing it. More than that, many of these stars must have been the high-mass variety which explode as supernovae. Chapter 6 described how supernovae are so bright that they can be seen across most of the visible Universe; so why can we not see these supernovae exploding? Quasars have been found with a redshift of 5 which (depending upon the choice of Hubble constant) corresponds to a universal age of just 1–2 billion years. Some astronomers believe that this leaves precious little time for the largely isotropic Universe of 300,000 years to fragment and gravitationally collapse into galaxies (as we shall discuss in the Chapter 9), let alone for generations of high-mass stars to live and die.

Terlevich began to run simulations in an attempt to determine the observational properties of such a bout of early star-formation. His conclusions echo the earlier theories about active galaxies and the possibility of their being powered by starbursts. He especially developed his calculation to account for the events towards the end of the starburst galaxy's life, when the massive stars begin to explode as supernovae.

The modern version of the starburst model can actually be split into four distinct phases of activity. In the first – which lasts for approximately three million years – the central region of the forming galaxy becomes an intensely bright emission nebula – similar to (but on a much larger scale than) the emission nebulae in our Galaxy. This has happened because the starburst has caused thousands upon thousands of stars to form with a range of all possible masses.

In the second phase of activity, the most massive of the stars – of 40–60 M_{Sun} – become Wolf–Rayet stars. These are very hot stars which shine between 100,000 and one million times more brightly as the Sun, and produce an incredibly large amount of dust in their outer atmospheres. One of the characteristics of a Wolf–Rayet star is that it sheds the outer layers of its atmosphere by expelling a shell of hydrogen. According to calculations associated with this model, the galaxy will now begin to resemble a Seyfert Type 2 with narrow hydrogen emission lines, because the hydrogen envelopes are only expanding slowly.

The third phase of activity begins when the starburst is 4–8 million years old. The massive stars now explode as supernovae. Although these explosions are titanic, they are relatively few in number, and the galaxy retains its Seyfert Type 2 appearance. The central regions are seeded with the heavy elements. The expanding supernova shock waves also create the low-level radio emission detected from radio-quiet AGN.

Finally, the fourth and most dramatic phase of activity begins when the starburst is more than 8 million years old. As time has gone by, so the lower-mass stars have begun to reach the ends of their lives. By about this stage in the starburst's evolution, stars of 8–25 M_{Sun} are beginning to explode as supernovae. There are many more of these stars than there are larger ones, and the effect of the explosions on the nucleus of the galaxy becomes very noticeable. With the violently ejected gas spreading all over the nucleus, the emission lines become Doppler broadened and begin to resemble Seyfert Type 1 or quasar line spectra. This type of activity, once it begins, is calculated to last about 50 million years.

This starburst model is applicable only to those AGN which are radio-quiet, but this is a staggering 99 in every 100. Even then, many astronomers have trouble accepting that this type of activity can persist for long enough to generate the luminosity displayed by the most powerful quasars. One of the most interesting, and totally independent, observations which increases confidence in starbursts was reported by Alex Fillipenko, of the University of California at Berkeley. His optical observations of the supernova which exploded in the spiral galaxy NGC 4615 during May 1987 produced some unexpected line profiles. He ascribed the profiles as being peculiar due to the dusty environment in which the star actually exploded; but he also asserted that if the supernova had taken place in the central region of the galaxy, it could easily have been misclassified as a Seyfert galaxy nucleus. It is obvious that to ignore the starburst model is a big mistake; it must have a place somewhere in explaining the grand cornucopia of galaxy types. The abundance of heavy elements in quasars, the ubiquitous nature of the IRAS galaxies, even the simple fact that galaxies have stars in them, point to star formation and starbursts as being of intense cosmological interest. For instance, we still do not know when, in the life of a galaxy, the first stars form.

An increasing number of astronomers have, with further study, begun to regard some of the ultraluminous IRAS galaxies as being a combination of starburst activity and active galactic nuclei. 324 IRAS galaxies are classed as ultraluminous because they have far-infrared luminosities which exceed 10^{12} solar luminosities. Study of these objects has shown that they all present evidence of interaction and merging. Although this will inevitably cause an intense starburst, many astronomers maintain that when the emission at different wavelengths is compared, the 'colour' of the galaxy is more like that of an AGN than of a starburst. Some studies of the ultraluminous IRAS galaxies Arp 220 and Markarian 231 have also revealed intensely compact infrared emission sources which are probably too small to be starbursts.

The continual study means that astronomers are beginning to realise that active galaxies and other peculiar galaxies are perhaps not as simple as was once thought. There are now several research groups who postulate that the central engine of an active galaxy is a black hole surrounded by an accretion disc, a BLR, a dusty torus and a NLR, which is then surrounded by a starburst. As well as the two previously mentioned ultraluminous IRAS galaxies, IRAS F10214+4725 is an excellent example. Certain characteristics of its spectrum indicate that it is a starburst, whilst other characteristics strongly indicate the presence of a black hole. The object has been simulated by David Clements, of the University of Oxford, using one of the hybrid 'dust enshrouded black hole surrounded by a starburst' models!

7.9   GAMMA-RAY BURSTS

A new perspective on the life (or, more accurately, the death) of stars in the early Universe has been provided recently by a new generation of gamma-ray detectors that are capable of locating the positions of bursts of high-energy radiation.

Gamma-ray bursts were discovered by the United States military in 1967. The chain of events that led to this event began in the 1950s, at the height of the United States' nuclear bomb testing programme and at the beginning of the space race. It was considered that space would be a good place in which to test nuclear weapons. The United States began two projects: the first was to test a nuclear weapon in space, and the second was to build a system capable of detecting whether any other nation (specifically Russia) had exploded a similar bomb in orbit.

In 1962, a project code-named Starfish successfully delivered and exploded a nuclear bomb in Earth orbit. The high-energy particles and radiation that the bomb generated – specifically gamma-rays and X-rays – caused a number of nearby satellites to malfunction, and some stopped working completely. This raised eyebrows in the US military, and soon the Squanto Terror programme was set in motion. In the event of a war, this would deliver atom bombs into orbit, and explode them next to enemy satellites to eliminate them.

Shortly after this, however, the Nuclear Test Ban Treaty was signed. This specifically forbade the detonation of nuclear weapons in space, and Squanto Terror was abandoned. The second project – to detect nuclear detonations in space – continued, as it could be used to enforce the Treaty.

In 1963, the first two satellites of the project were launched. Vela 1 and Vela 2 were 20-sided polyhedrons able to detect X-rays or gamma-rays emanating from nuclear explosions in space. The Vela programme continued until 1970, during which time a total of 12 increasingly sophisticated and sensitive Vela satellites were launched.

In 1967, the Vela satellites recorded a burst of gamma-ray activity coming not from Earth orbit but from deep space. It lasted for six seconds. By the mid-1980s several hundred gamma-ray bursts had been detected. Astronomers were presented with a major puzzle, as no two gamma-ray bursts were the same. Some bursts displayed very high energy and lasted for a few seconds, and others were less bright but persisted for longer, providing astronomers with a major clue that the events occurred far outside the Galaxy; the rays from the longer, dimmer bursts had been 'stretched' by the expansion of the Universe. However, some astronomers were not convinced that the events were as violent as they seemed, and so felt they were probably taking place closer, in the near vicinity of the Milky Way.

In 1991 NASA launched the Compton Gamma-Ray Observatory from the Space Shuttle *Atlantis*. It was a major step forward in the study of gamma-ray bursts, since it could locate their positions. Because gamma-rays are difficult to reflect or refract, it had until then been impossible to obtain images of them. The Compton Observatory, however, was equipped with eight detectors which could be used together to determine the direction from which the bursts were coming. It was swiftly realised that these events took place every day and came from completely random directions in the sky.

The puzzle remained. With only a few seconds-worth of gamma-rays to work with, astronomers could let their imaginations run wild about what was causing the bursts. They wondered about neutron stars either colliding or suffering starquakes; they discussed exploding black holes; and they considered active galactic nuclei.

Everything changed in 1996, however, with the launch of the Italian–Dutch BeppoSAX satellite. This observatory could swiftly determine the location of a gamma-ray burst and communicate the information to astronomers all over the world. If those astronomers were quick enough, they could then point their telescopes in the direction of the burst, hoping to see the optical counterpart – the visible radiation associated with the burst.

On 28 February 1997, BeppoSAX detected a burst, which was designated GRB970228. At the William Herschel Telescope on La Palma, two optical images were taken – the first just 21 hours after the burst's detection and the second a week later. A comparison of the images showed that a light-source on the first image did not appear on the second. The optical counterpart had been found. Now, the real work could begin. On 13 March the New Technology Telescope, in Chile, took an image of the gamma-ray burst's field and discovered a very faint galaxy in its location. On 26 March and 7 April the Hubble Space Telescope obtained deeper

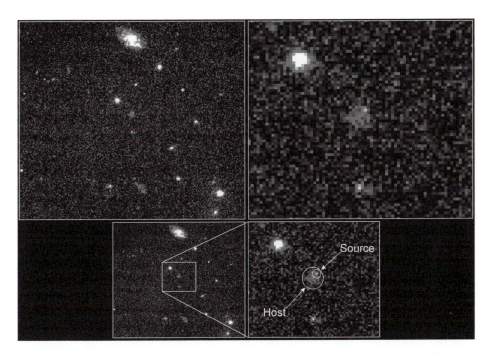

Figure 7.16. In this Hubble Space Telescope image of the gamma-ray burst GRB970228, the optical counterpart is seen at the centre of the image (top left) and in the centre of the expanded view (top right). Analysis of the image shows that the source location is not in the centre of the host galaxy (bottom). (Photograph reproduced courtesy Andrew Fruchter (STScI), Elena Pian (ITSRE-CNR), and NASA.)

images which showed that the very faint gamma-ray burst's optical counterpart had, in fact, not yet faded to complete obscurity, and that it was offset from the centre of the newly discovered galaxy (see Figure 7.16).

A spectrum of the fading burst showed that it was located at a redshift of 0.835. – approximately halfway across the observable Universe. Because it was located away from the centre of a galaxy, the active galactic nucleus idea was discounted. One interesting theory is that a gamma-ray burst might be the emission from a very high-mass star in which the core has collapsed into a black hole rather than becoming a neutron star, which is more usual. When a core becomes a neutron star, stellar material rains down and collides with it, setting up a shock wave which travels through the star, resulting in a supernova. If the core becomes a black hole, however, the raining material will simply be sucked out of existence. The appearance of a black hole in the centre of the star swiftly liberates a colossal amount of energy from the surrounding gas falling into its gravitational grip. Such an event has been termed a 'hypernova', and is perhaps, at the moment, the leading explanation for gamma-ray bursts.

The study of optical counterparts is now advancing rapidly. A burst which took place on 14 December 1997 (GRB971214), has astounded (and confounded)

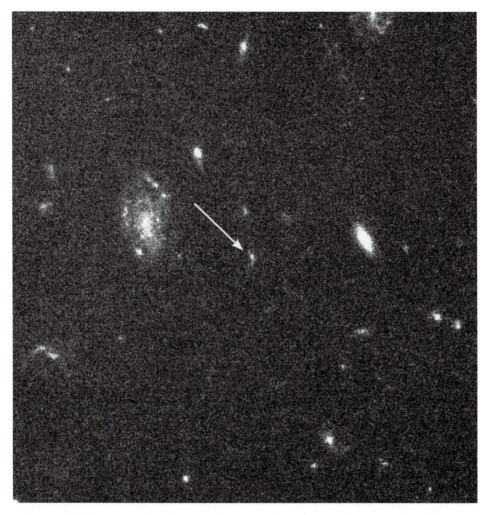

Figure 7.17. Four months after the discovery of gamma-ray burst GRB971214, the Hubble Space Telescope obtained this image of its faint host galaxy. The HST's data have helped to confirm a previous observation of this galaxy, made by one of the Keck 10-m telescopes, that places its distance at 12 billion light years. (Photograph reproduced courtesy S. R. Kulkarni and S. G. Djorgovski (Caltech), the Caltech GRB Team, and NASA.)

astronomers throughout the world (see Figure 7.17). Its redshift of 3.42 indicates that this event took place about two billion years after the Big Bang. The analysis of its optical counterpart suggests a minimum energy for this event such as may be expected from a hypernova. The analysis of the gamma-rays themselves suggests something unprecedented.

According to Caltech professor Shrinivas Kulkarni and his team, the energy released by this event is several hundred times greater than a hypernova could generate. In fact, the energy involved in this one single event must have created conditions in its near vicinity similar to those thought to be found in the Universe just one millisecond after the Big Bang.

Another monstrous gamma-ray burst took place on 23 January 1999. It was discovered by the Compton Gamma-Ray Observatory, which transmitted approximate rough coordinates to the Gamma-Ray Burst Co-ordinates Network (GCN), based at NASA's Goddard Space Flight Center in Greenbelt, Maryland. The GCN immediately forwarded the position to observatories around the world. The Robotic Optical Transient Search Experiment (ROTSE) in Los Alamos was the first facility to respond, and took pictures of the region of sky where the burst was detected. The first picture showed a new star within the region where the explosion was reported. It was becoming brighter, and achieved peak brightness just five seconds after ROTSE began observing it.

The burst was also picked up by the BeppoSAX satellite. Within three hours of the gamma-ray burst, the Kulkarni team obtained images of the fading optical counterpart, using the 60-inch Mount Palomar telescope. The next night, a team of astronomers led by Dr D. Kelson, of the Carnegie Institution of Washington, used the Keck II 10-m telescope to measure the redshift of the burst's host galaxy. They found it to be at a distance of nine billion light years.

The final stage of this astronomical detective hunt was completed on 8 and 9 February when the Hubble Space Telescope imaged the fading burst and its host galaxy. By then, the burst's optical emission had dimmed to less than four millionths of its original brightness. The total energy that the burst radiated into space was the equivalent of the radiation produced by a staggering 100 million billion stars: the stellar content of a million galaxies (see Figure 7.18).

As yet, no convincing theory to explain the violence of these gamma-ray bursts has been forthcoming. There also exists the ever-present possibility of faulty assumptions in determining the exact energy of the event, which may have raised it beyond acceptable levels. However, in the near future, BeppoSAX will continue the search; and the 1999 launch of NASA's High Energy Transient Experiment II (HETE II) will be followed in 2005 with the Gamma-Ray Large Area Space Telescope (GLAST), which will search for high-energy gamma-ray bursts similar to GRB971214.

7.10 THE EVOLUTION OF ACTIVE GALAXIES

The importance of active galaxies in the study of the younger Universe cannot be overstated. Even the Hubble Space Telescope struggles to observe galaxies at redshifts greater than 1, whilst quasars burn brilliantly out to redshifts of 5. One particular observational oddity of quasars is that their density appears to peak at about redshift 2. This oddity was first pointed out by Geoffrey Burbidge in 1967. At first it was not known whether this peak was real or the effect of our current selection criteria and state of technology. It now appears that the peak is real, and provides strong evidence for the evolutionary nature of the Universe and the celestial objects

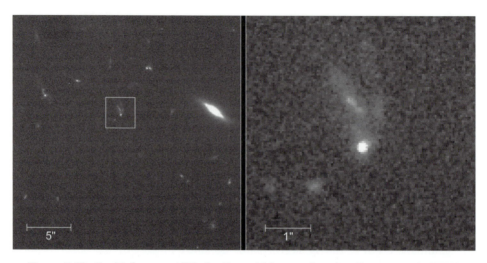

Figure 7.18. On 23 January 1999 the Space Telescope Imaging Spectrograph (STIS) imaged the gamma-ray burst GRB990123. The images here were obtained on 8 and 9 February, by which time the burst had faded to 4 million times less than the brightness it had attained. Even so, it is still brighter than its host galaxy – a strange finger-like projection that is not a classic spiral galaxy nor an elliptical galaxy. It is almost certainly a starburst galaxy located at a great distance. (Photograph reproduced courtesy Andrew Fruchter (STScI) and NASA.)

within it. (The reasons for this peak will be explored in the next chapter.) In fact, they point the way towards a more complete understanding of how the IRAS galaxies and the active galaxies are related to the normal galaxies. Tentative answers can now be provided to such questions as: when does a normal galaxy become normal? Are some galaxies born normal and others born active, or do all normal galaxies go through an active stage as part of their evolution? Can dormant active cores be turned on again?

In order to answer these questions we have to take one more step towards the edge of the Universe, and journey into the astronomical wilderness where galaxies form.

8

Large-scale structures

The limits of our technology can, if we are not careful, present us with a distorted view of the Universe. Our detection of normal galaxies has been confined to a volume of space out to a redshift of approximately 0.1. Traditionally, galaxies detected at this limit were termed high redshift galaxies but, essentially, were no different from others of their kind at lower redshifts. The more distant Universe, as we described in the previous chapter, was the realm of the active galaxies. Although examples of these are present in the modern Universe – at redshifts comparable with normal galaxies – the vast majority are recognised only far away in the distant realms of the cosmos out to redshift 5. The reason why these active galaxies have been thought to dominate this realm is because technology has been too limited to be able to detect anything as faint as a normal galaxy; but with the increase in both the sensitivity of CCD cameras and the sizes of the largest telescopes in the world, the situation has now begun to change. The Hubble Space Telescope is also making a valuable contribution to this study.

8.1 CLUSTERS OF GALAXIES

The Universe is finite in age and, therefore, at some stage in its past the galaxies must have formed. From the observations of the Universe we have made so far, it seems to be a reasonable assumption that the epoch of galaxy formation occurred between redshift 5 and redshift 1000. The former boundary is rather arbitrarily chosen because it marks the position at which our current technology fails to detect active galaxies. The latter boundary is well chosen, however, because this is the point at which the cosmic microwave background radiation has been detected.

Our first step towards detecting the formation of galaxies is to try to observe their evolutionary progress. When Edwin Hubble originally classified the galaxies according to their morphological type, he formed the opinion that his tuning fork diagram displayed an evolutionary path. In his view, galaxies began as spherical conglomerations of stars which, over the course of time, gradually flattened out to become the more eccentric ellipticals. He thought that these rotating systems, having flattened out, then developed spiral arms and became either spiral or barred spiral galaxies, depending upon the exact details of their internal structure. Despite the simple elegance of this idea, it is wrong. Galaxy evolution is a much more complex and subtle process which we are only now beginning to understand.

An advantage in this quest is that the galaxies are grouped into clusters, which presents us with ready-made laboratory samples at different redshifts. (It is important to remember that the redshift scale is not actually linear; even though redshift 0.1 would appear to seem very close by, it actually encompasses a colossal volume of space (see Figure 8.1)). This is analogous with star clusters, of which there are two distinct types: open clusters and globular clusters. Both contain stars yet, in our overall understanding of the cosmos they are very different from one another in terms of age, evolutionary history and position. All of this has been discovered simply by taking that first step and classifying the star clusters by their morphology. Similarly, the histories of galaxies can be inferred depending upon whether the system is spiral or elliptical (as discussed later in this chapter). With this in mind, the classification of a galaxy cluster, based upon its shape and its contents, may be the first step towards understanding how it, and the galaxies within it, formed. In the 1950s, two astronomers began to compile independent catalogues of galaxy clusters. George Abell took as his definition of a cluster only the brightest and densest of

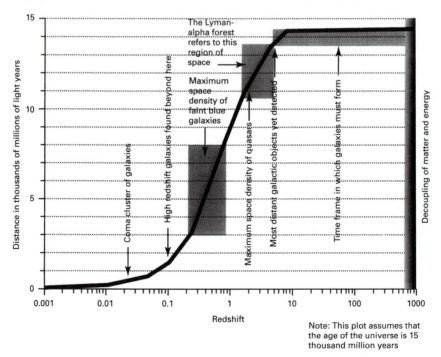

Figure 8.1. As studied by Hubble and Humason, the redshift of a galaxy appears to be linearly proportional to its distance. This diagram shows the relationship.

Figure 8.2. The Virgo cluster contains mostly spiral galaxies, although the central region is dominated by elliptical galaxies. (Photograph reproduced courtesy of AURA/ NOAO/NSF.)

aggregates, whilst Fritz Zwicky set about defining every conglomeration no matter how small or how sparse. Consequently, Abell's catalogue of rich clusters, published in 1958, contained 2,712 members, whilst Zwicky's more comprehensive catalogue listed 9,134 galaxy clusters and was published over an eight-year period from 1960 to 1968. In the course of their efforts, both astronomers realised that galaxy clusters can be classified according to their appearance. Abell thought his bifurcated into regular and irregular clusters, whilst Zwicky felt that a better system was to classify according to the density of galaxies – compact, medium or open clusters.

Virtually concurrent with this work, William Morgan was studying the contents of galaxy clusters. He determined that cluster type could be defined according to which Hubble type of galaxy was predominant. His initial work allowed him to define two types of cluster. The Virgo cluster type (after which the prototype was named) is one in which the majority of galaxies are spirals (see Figure 8.2). Only a few of the very brightest members of this type are ellipticals or lenticular galaxies. The brightest are concentrated towards the central regions of the cluster. Morgan's second type took as its prototype the Coma cluster (see Figure 8.3). In examples of these, ellipticals and lenticulars predominate among the brighter cluster members. What few spirals there are, tend to be confined to the much dimmer cluster members. According to Morgan's studies, the majority of nearby clusters are of the Coma type.

Morgan continued his work on defining the clusters by examining the galaxies they contain. In 1964 he showed, with the help of Thomas Matthews and Maartin Schmidt, that a relatively common feature in some clusters is the presence of one

Figure 8.3. In contrast to the Virgo cluster, the Coma cluster is dominated by elliptical galaxies, and there are few spiral galaxies. Most nearby clusters are of the Coma type. (Photograph reproduced courtesy of AURA/NOAO/NSF.)

exceptionally large, luminous galaxy at the centre of the cluster. They are so large and diffuse, however, that despite their relatively large luminosity they actually display a low surface brightness. They were termed cD galaxies (D denotes diffuse). The 'c' was used to indicate the luminous nature of the object, in the same way that this letter was sometimes used as a spectroscopic designation for superluminous stars, although many simply remember it as denoting a 'cluster-dominating' galaxy. By continuing to analyse the ubiquity of this type of galaxy, Morgan and his new co-worker, Laura Bautz, introduced a classification scheme for clusters which relied upon the presence or absence of bright cluster-dominating galaxies: a Type I cluster includes a single cD galaxy or a giant binary galaxy resembling a dumbbell; a Type II cluster is dominated by two or three bright galaxies; and a Type III cluster has no particularly dominant galaxy.

8.2   THE ROOD–SASTRY CLASSIFICATION SCHEME FOR GALAXY CLUSTERS

Three different ways of classifying clusters had therefore been devised: the two morphological systems of Abell and Zwicky, and the contents criteria used by Morgan and Bautz. What was needed was an interpolation between the three to determine whether cluster morphology influenced cluster contents. This work was performed by Herbert J. Rood and Gummuluru Sastry, and was presented in 1971. In an attempt to reconcile the two approaches, Rood and Sastry analysed the spatial arrangement of the brightest 10–20 members of the cluster and the predominant type of galaxy. By doing so, they could identify six types of cluster (see Figure 8.4).

Type cD was retained, the definition in the Rood–Sastry system being that a cD cluster is dominated by a single large cD galaxy, a slightly smaller and less luminous D galaxy, or a giant elliptical galaxy. Whatever the precise classification of the

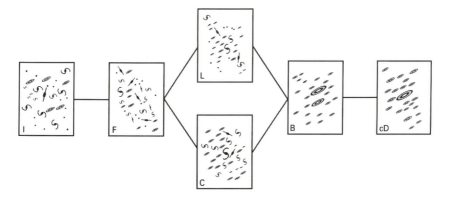

Figure 8.4. The Rood–Sastry classification scheme provides a broad evolutionary sequence for the dynamical development of galaxy clusters.

central galaxy, in order for the cluster to be of this type its central galaxy must be at least three times larger than its nearest rival. Three sub-groups are also possible, based upon the exact appearance of the cD galaxy: a cD_s galaxy has a satellite galaxy visible within its faint, outer envelope of stars; a cD_p galaxy has a peculiar feature, such as a jet or a tail or a skewed nucleus or envelope; and a cD_n galaxy displays multiple nuclei, although care must be taken to make sure that the observation has not been the result of the intrusion of foreground galaxies. The members of a cD cluster are mainly elliptical and lenticular galaxies.

Type B clusters are dominated by a pair of galaxies, both of which are at least one magnitude brighter than any other cluster member. They must also be separated by less than ten times the diameter of either galaxy. In the case of the sub-group, B_b, they are joined together by a luminous bridge of material, or even share a common envelope of stars. This latter situation matches the observation of the dumbbell-shaped galaxies noted in the Morgan–Bautz classification scheme.

A Type C (core-halo) cluster consists of several bright galaxies of any type which constitute the core, surrounded by many fainter galaxies which form the halo.

Type L galaxy clusters are similar to Type C, except that instead of the bright galaxies being confined to a core, they are spread out in a line. The fainter galaxies are sometimes found to lie within extensions of these lines, whilst in other examples they simply surround the bright galaxies. A subgroup is denoted L_a, which means that the bright galaxies lie in an arc rather than a straight line.

Type F clusters appear flattened or elongated on the sky. They are not dominated by any bright galaxies, however, and the cluster members are predominantly spirals and lenticular galaxies.

The final class in the Rood–Sastry scheme is Type I – irregular. These clusters are not dominated by any galaxy, and appear irregularly spread across the sky with no central concentration. In examples of this kind, almost every single cluster member is a spiral galaxy. Sub-groups are I_c – clusters which contain sub-groupings of galaxies throughout their volume – and I_s – clusters which show no concentrations whatsoever.

Having identified these six different types of galaxy cluster, the first thing to be noticed about them is that the morphology affects the content. For example, cluster types B and cD are predominantly composed of elliptical galaxies, whilst I and F clusters contain mainly spiral galaxies. C and L clusters appear to be intermediate in galaxy population between the two other pairs; that is, they have almost equal numbers of spirals and ellipticals. The difference in galaxy type also means that clusters also contain different stellar populations. Ellipticals – and therefore the clusters in which they predominate – contain old stars, whilst spiral rich clusters are populated by young stars. Is this a clue that perhaps the type of a cluster signifies its age? If we assume, for a moment, that I and F clusters are the youngest, and that B and cD are the oldest, then it would therefore seem logical to place them in an evolutionary sequence.

This evolutionary picture has been broadly substantiated by both observation and theoretical modelling, but with one major caveat. The Rood–Sastry type is not simply an indicator of the cluster's chronological age; there is also a strong evolutionary component which depends upon the cluster's density. We shall return to the age of the clusters in due course, but for the moment we shall concentrate upon their evolution. It was stated earlier that many of the nearby clusters are dominated by elliptical galaxies, which would provide our first corroborating evidence for this idea. A number of observations of galaxy interaction and mergers show that spiral galaxies lose their distinctive arms in close encounters with one another, and as they gradually merge together they take on the appearance of an elliptical galaxy. The process strips the dust and gas from the spirals by throwing it into intergalactic space. This effectively halts star formation within the merged pair once the starburst, triggered by the merger, subsides. Interestingly, the egress of dust and gas can also trigger dwarf galaxy formation in the outflow streams.

If elliptical galaxies are forged in mergers, then the Rood–Sastry classification must also be telling us about the density and clumpiness of a galaxy cluster. 'Clumpiness' does not refer to a cluster which has a central concentration; it means quite the opposite. A clumpy cluster demonstrates an open distribution of galaxies; that is, the galaxies appear to be randomly clumped throughout the cluster's volume. cD and L clusters are the densest and least clumpy, whilst B, C, F and I are less dense but clumpy. This behaviour suggests that the densest galaxy clusters are dynamically the oldest, whilst the less dense galaxies are dynamically younger. 'Dynamically old' means that the constituent galaxies have had plenty of opportunities to interact with one another. This process allows for mergers and collisions whilst dissipating a lot of the galaxies' kinetic energy. Thus, the cluster becomes more concentrated towards the centre of mass. A dynamically young galaxy is one in which there has been little opportunity for the member galaxies to interact and merge with one another. This might be because the cluster is chronologically younger in age, thus reducing the time available for interactions. It may also be less densely packed, thus cutting down the opportunities for interactions.

Computer simulations have shown that, by starting with an irregularly shaped cluster, in which galaxies are distributed at random (a Type I cluster), the system can evolve through gravitational interactions into denser and denser types of clusters. As it does so, the population of elliptical galaxies grows whilst the number of spirals dwindles. The cD galaxies, which are always found at, or very close to, the centre of mass of clusters, are thought to have grown to such huge sizes because they have repeatedly cannibalised smaller galaxies. These examples of the multiple nuclei cD galaxies would appear to have been caught in the devouring process!

Are irregular, clumpy spiral rich clusters, therefore, really younger than the denser, elliptical rich collections? Although it is true that clusters at higher redshift appear to contain a larger proportion of spiral galaxies, the available evidence appears to disavow this simple interpretation. Cluster formation – rather like galaxy formation – is something that has yet to be conclusively observed. One of the questions that we will later consider is whether clusters formed first and then fragmented into individual galaxies, or whether fully formed galaxies later aggregated into clusters. Whatever the precise formation scenario, it would appear that the majority of a cluster's aggregation of mass has taken place by a redshift of 5. In essence, galaxy cluster formation takes place within a few thousand million years of the birth of the Universe. So, if the clusters are all approximately the same age, then the Rood–Sastry classification must also be a pointer to the initial conditions of the cluster. This is because we have stated that the other factor, which has the most effect on the dynamical evolution of a cluster, is its density. The greater the concentration of matter, the higher the likelihood of galaxy interactions and hence the faster the cluster will evolve into a Type cD or B.

Instead of indicating the age of the cluster, its Rood–Sastry type can actually be used to infer something that, cosmologically speaking, is far more important: we can gauge the initial density of that region of space. The three-dimensional distribution of galaxy clusters can therefore reveal the architecture of the cosmos at a time when the Universe was less than one sixth of its present age.

8.3   EVOLUTION OF GALAXY CLUSTERS

Although astronomers have yet to see the actual birth of a galaxy cluster, they have perhaps discovered the proverbial 'smoking gun' – an effect which takes place after the actual event, but provides a clue to how it all happened.

The discovery came in 1970, shortly after the launch of the first X-ray satellite, Uhuru. This was the first time that astronomers had really been able to study the Universe in X-rays, and it was a revelation. Edmin M. Kellogg and his team, from the hi-tech company American Science and Engineering, focused on the Virgo and Coma clusters of galaxies. They found that instead of discrete galaxies being attracted by their mutual force of gravity, the clusters appeared to be enormous 'balls' of super-hot gas with a typical temperature in the order of 25 million K. At that temperature, the gas would naturally emit copiously at X-ray wavelengths.

The amount of intergalactic gas in each cluster was usually as much, if not more than, the amount of matter contained in the galaxies themselves. It swiftly became obvious that a cluster of galaxies should really be thought of as a ball of hot gas in which individual galaxies are embedded like pips in a cosmic watermelon.

With the advent of more advanced X-ray satellites, the shapes of the gas clouds have become resolvable. Interestingly, the Coma cluster appears mostly regular, with smaller gas clumps inside (see Figure 8.5), while the Virgo cluster is more irregular in shape (see Figure 8.6). The small gas-clumps inside the Coma cluster are thought to be smaller groups of galaxies that are merging with the much larger cluster. The irregular shape of the Virgo cluster, however, is thought to be caused by its having yet to finish pulling in matter from its surroundings.

The Virgo cluster is still forming from very large-scale gravitational collapse, whereas the Coma cluster is more or less formed, but has been caught in the act of swallowing a few small groups. Galaxy clusters can therefore form either from the merger of smaller groups, or by the large-scale gravitational collapse of matter. Alternatively, perhaps any individual cluster has undergone both processes in its long life.

The clusters in the nearby Universe are now relatively quiescent, even though astronomers have discovered this dichotomy of activity in the Coma and Virgo

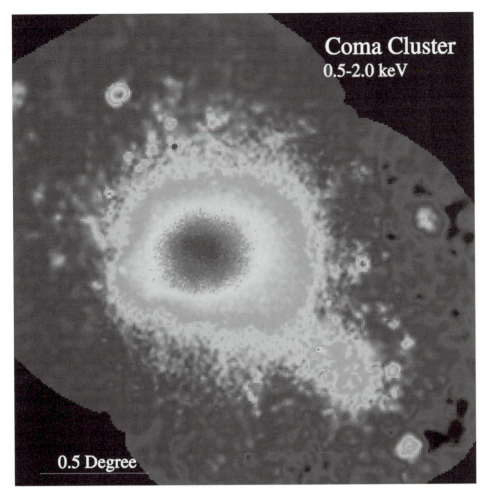

Figure 8.5. X-ray emission from the intergalactic dust in the Coma cluster. The individual galaxies do not appear in this image. (Photograph reproduced courtesy of S. Snowden and NASA.)

clusters. It can therefore be assumed that cluster growth has all but stopped in the nearby Universe. Looking back with better and better telescopes into the Universe's past, however, it is hoped that more violent formation and merger activity will eventually be seen.

8.4   THE RECOGNITION OF LARGE-SCALE STRUCTURE

The largest structures in the Universe are composed of galaxy clusters. Chapter 1 mentioned Charles Messier, who was the first person to begin cataloguing galaxies (or nebulae, as they were known then). Messier had catalogued them not because he

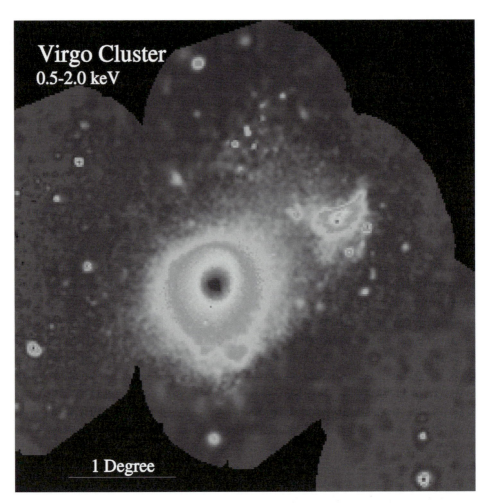

Figure 8.6. X-ray emission from the Virgo cluster. As with the Coma cluster (Figure 8.5), the X-ray emission comes from the intergalactic dust, not from the individual galaxies. (Photograph reproduced courtesy of S. Snowden and NASA.)

thought they were interesting, but rather because he wanted to avoid them. He even went so far as to call these fuzzy patches of light 'the vermin of the sky'. He was a comet hunter, and compiled his catalogue so that if he suspected that he had discovered a comet, he could check whether he had instead encountered one of these static nebulae.

In 1783, William Herschel began studying the skies with a view to cataloguing the locations of these nebulae. Using state-of-the-art equipment, he discovered 2,500 nebulae in the northern skies, the majority of which we now know as galaxies. His son, John Herschel, continued the work, re-observing the northern skies, confirming his father's discoveries, and finding an additional 500 nebulae.

John Herschel's next move was a bold one. He relocated his telescope to the Cape of Good Hope in the southern hemisphere. This was no mean feat, because although his telescope was modest in aperture by today's standards – 18 inches – its focal length was 20 feet, making it monstrously long. It was supported by a triangular wooden structure which pointed the telescope at the sky.

Herschel spent four years in the southern hemisphere, cataloguing the skies and discovering a further 1,700 nebulae. Eventually, in 1864 he published a compilation of all known nebulae. It was called the *General Catalogue*, and it is still in use today, under the guise of the *New General Catalogue*, following its recompilation and enlargement by J.L.E. Dreyer, Director of Armagh Observatory, in 1888. Many galaxies and nebulae are still known by their NGC numbers.

It was while compiling the *General Catalogue* that John Herschel recognised signs of a truly large-scale structure in the unknown nebulae he was discovering. He once commented that a third of the nebulae he was studying were located in just one eighth of the sky. That region was in the constellation of Virgo. The first thoughts were being implanted in astronomers' minds that some of the nebulae in the sky were distant stellar systems, similar in structure to our own Milky Way. It was known as the 'island Universe' theory, and Herschel spent some of his life believing this point of view. During this time he described the way in which other concentrations of galaxies seemed to congregate around Virgo. He even went so far as to suggest that the Milky Way was involved in the congregation, but lay on its outer fringes. This was the first recognition of the large-scale structure we know today as a supercluster, and Herschel's description of what we now call the Local Supercluster was remarkably accurate (see Figure 1.8). In more recent times, Clyde Tombaugh championed this point of view and helped to bring about the recognition of such large-scale structures.

By the careful study of galaxies' redshifts, large-scale maps of the Universe have revealed glimpses of truly staggering structures. Stretching through space there are chains, filaments and walls of galaxy clusters, surrounding enormous voids in which the number of galaxies drops so dramatically that compared with the rest of the Universe they are as good as empty.

In our 'local' Universe – defined as a sphere of radius 2.5×10^8 light years, centred on the Local Group of galaxies – the Virgo cluster and its associated supercluster are the dominant features. The Virgo cluster itself contains material equivalent to about 5×10^{14} M_{Sun}. Amazingly, almost a tenth of this mass lies in the gigantic elliptical galaxy M87 (see Figure 8.7). Elliptical galaxies in the cluster number about 80, whilst the spiral galaxies are roughly one-and-a-half times as numerous and appear to cluster around the ellipticals, forming a sort of core-halo arrangement. The most numerous are the dwarf galaxies, of which around 900 have been identified; and there may be many more which are simply too faint to be seen.

The Virgo supercluster connects other groupings of galaxies, such as the Local Group, to the Virgo cluster via 'appendages' of galaxies that weave their way out into the surrounding space. The radius of the supercluster is estimated to be slightly less than 5×10^7 light years.

As large as this structure is, it appears to be no more than a component in the great Centaurus Wall of galaxies. It is a conspicuous feature of galaxy maps, and is

Figure 8.7. M87 is the dominant galaxy in the nearby Virgo cluster of galaxies. Although the Virgo cluster contains more than 1,000 galaxies (mostly dwarf galaxies), M87 contains one-tenth of the mass of the cluster. (Photograph reproduced courtesy AURA/NOAO/NSF.)

estimated to have dimensions of $3.5 \times 10^9 \times 2.5 \times 10^9 \times 5 \times 10^7$ light years. There are a few very large clusters of galaxies in the Centaurus Wall, such as Abell 3627. Perhaps the most interesting aspect of the Centaurus Wall is that the galaxies in our local supercluster are moving towards it. This has led to the wall being termed the 'Great Attractor'.

Other structures similar to the Centaurus Wall in the local Universe have been identified, and have been named the Hydra Wall, the Fornax Wall, the Cetus Wall and the Coma Wall. The Coma Wall – the first such structure to be identified – is centred on the Coma cluster of galaxies, and was dubbed the Great Wall before the frequency of these structures was known.

There are also voids in the local Universe. The nearest of them is called the Local Void, with the Virgo supercluster and the Centaurus Wall appearing to run around some of its boundary. The void covers nearly one third of the sky but, in reality, it is not very large. The Microscopium Void is much larger, and the Eridanus Void is a rambling, elongated void that appears to be a merger of other, smaller, spherical voids.

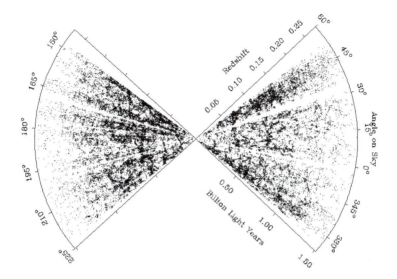

Figure 8.8. 2dF redshift map. At the time of writing this is the largest map of the Universe yet recorded. 49,636 galaxies are plotted by their redshift and direction from Earth. The 2dF continues the work of collecting nearly eight times as many galaxy redshifts. (Map reproduced courtesy Anglo-Australian Observatory.)

On the whole, the large-scale structure of the local Universe is like a mass of interconnected bubbles. The volumes inside the bubbles are the voids, and the galaxy collections are clustered on the surface of the bubbles.

The biggest map of the Universe to date was released in March 1999 by the scientists involved in the 2dF (2-degree Field) project at the Anglo-Australian Observatory (see Figure 8.8). It plots 49,636 galaxies, and places the project well on its target of 250,000 galaxies by the year 2001.

8.5   THE LYMAN-α FOREST

As well as studying the three-dimensional structure of galaxies in the Universe, astronomers would like to study the diffuse gas which has not yet collected into galaxies and transformed into stars; and there is an ingenious method of doing so which is made possible because of the structure of the hydrogen atom.

The spectrum of the hydrogen atom was investigated by the American physicist Theodore Lyman during the first two decades of the twentieth century. Specifically, he investigated the electron transitions which ended or originated at the first energy level. The brightest of these became known as the Lyman-α line, which occurs at a wavelength of 122 nm in the ultraviolet. The value of the Lyman-α line soon became apparent after the discovery of quasars in the 1960s. Because of their large redshifts, the light we receive from them at visible wavelengths has usually originated in the ultraviolet region of the spectrum. The Lyman-α emission line is visible in the optical part of the spectrum of any quasar with a redshift greater than 1.7. Observations of these spectra soon revealed another interesting phenomenon. Shortward of the (usually) large Lyman-α emission line is a plethora of absorption lines. This forest of lines cannot be due to absorption in the quasar itself, because they are all Lyman-α absorption lines but spread across a range of successively lower redshifts. Instead, it is thought that this absorption originates in intervening clouds of hydrogen gas which exist along the line of sight between ourselves and the quasar under observation (see Figure 8.9). This being the case, the Lyman-α forest in a spectrum can be a very sensitive indicator of the distribution of hydrogen gas in the direction of the quasar. It can be used to determine the redshifts at which these clouds exist and, in certain cases where the lines are well defined, it can be used to estimate some of a cloud's properties: for example, its mass. The ozone layer of the Earth limits us to observations of the Lyman-α lines which originate in clouds at redshifts smaller than 1.7. At this redshift, the wavelengths of the absorption lines are too short to be able to pass through the Earth's atmosphere. The upper limit of these observations is obviously dictated by the distances of the individual quasars being studied.

Analysis suggests that hydrogen clouds are more massive with increasing redshift – presumably because less of the gas has been subsumed into stars and galaxies. The enrichment of the intergalactic medium with heavy elements appears to have been an early event. As mentioned in Chapter 7, quasars at high redshifts show lines due to metals in their spectra. Other metal absorption lines in spectra indicate that at least a generation of stars has already lived and died in the young Universe, which raises an interesting idea.

Traditionally, stars are divided into two generations: Population I and Population II. A Population II star is one which evolved when the Universe was young. It has a low metal content and is now very old – usually around 10,000 million years. A Population I star is a star which has formed subsequent to the enrichment of the Universe with metals from the dead Population II stars. Some astronomers favour the formation of a primordial population of stars which appeared very soon after the decoupling of matter and energy. These have been referred to as Population III stars, and would be composed of virtually nothing but hydrogen and helium.

Some of the high-redshift Lyman-α lines have been observed to have associated carbon features. This would appear to favour an interpretation in which a generation of stars has already lived and died in the early Universe and have seeded space with heavier elements. Another possible explanation is that the Lyman-α clouds themselves are present in star-bearing galaxies. Corroborating this view is the evidence that some of the Lyman-α lines display the characteristics of being produced by rotating clouds of gas, which might possibly be spiral galaxy discs in the process of formation.

At redshifts of less than 1.7, direct observations have shown that most of the gas which was once strewn throughout intergalactic space is now confined to large spherical haloes around galaxies. These haloes lie within a radius of 40 kpc from the galaxy's nucleus. The Universe has therefore changed from its almost homogeneous form – as indicated by the microwave background radiation – into a very unevenly

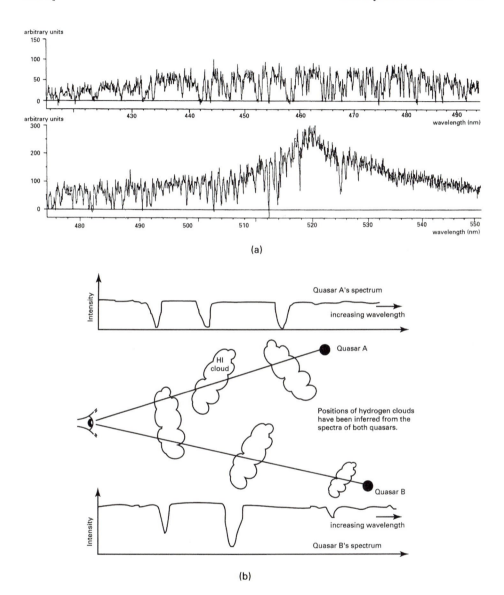

(a)

(b)

Figure 8.9. The Lyman-α forest. The optical spectrum of a quasar may exhibit a large number of absorption lines, in addition to emission lines. (a) In this spectrum of the quasar PKS 2126-158 can be seen dozens of narrow absorption lines attributed to the Lyman-α line of atomic hydrogen, comprising the Lyman-α forest. These are caused by multiple clouds of intergalactic hydrogen lying along the line of sight between the quasar and the observer. (Adapted from Audouze, J., and Israel, G., *Cambridge Atlas of Astronomy*, Cambridge University Press, 1998.) (b) The absorption lines appear strung out across the spectrum because the clouds are all at different distances.

distributed collection of mass (galaxies and clusters) which is homogeneous only on very large scales (see Chapter 4). These haloes of gaseous material appear to be almost totally independent of the properties of the galaxies which they surround. The only weak dependence is on the mass of the galaxy. This would seem to be clear evidence that the halo is produced by infalling gas, rather than outflow from the galaxy itself. This infalling of matter may be the final stages of galaxy formation itself, and an essential ingredient in keeping a galaxy fuelled with star-forming material.

8.6   THE EVOLUTION OF GALAXIES

Sooner or later, and, despite all the constraints and theories, the only way in which to truly understand galaxy formation is to observe it. In order to do this, astronomers must peer ever deeper into the cosmos, searching for ever fainter objects. As they have done this, so it has become obvious that the appearance of galaxies changes with redshift. In 1960, Rudolph Minkowski crowned his astronomical career with a discovery made on his last night observing with the 200-inch telescope at Mount Palomar. He discovered the most distant galaxy then known. It was the optical counterpart of the radio galaxy 3C295, and possessed a redshift of 0.46. Although this is not a large redshift by today's standards (radio galaxies are now known out to redshifts of 4.25), when compared with the Coma cluster at redshift 0.023, and with the Hercules cluster at redshift 0.036, it was a massive leap outwards.

The most obvious feature of 3C295 was that it was very red in colour. This is not too surprising, as elliptical galaxies, on the whole, contain old stars of spectral types G and K. These stars are yellowish-red in colour, and so the galaxies which contain them are also yellowish-red. The spectrum of 3C295 had been redshifted by 46%, and so had transformed most of its yellow light into red light. With little blue light to be redshifted into the yellow part of the spectrum, the galaxy therefore appeared as very red. Over a decade later, when astronomers had discovered a sufficient number of galaxies at this kind of redshift to be able to distinguish individual clusters, they began to apply corrections to the spectra to compensate for the cosmological redshift. To their surprise, many of these distant galaxies, including the ellipticals, appeared rather bluer than expected. Even the cluster containing 3C295 was shown to contain an abundance of blue galaxies. It appeared that these elliptical systems were not as old as those in the present-day cosmos. Some were obviously being caught in their final acts of star formation, as indicated by the blue light from young, massive stars.

Since then, the study of high-redshift clusters of galaxies has continued to advance with the invention of new technologies. Of the available images (at the time of writing), the best have been obtained by the Hubble Space Telescope. Ground-based observations are hampered by the atmosphere. The angular size of distant galaxies is so small that atmospheric effects wash out any detail. The only analysis possible has been photometric and spectroscopic.

The first tantalising images to be returned by the orbiting telescope were of the clusters CL 0939 + 4713 in Ursa Major and CL 0024 + 1654 in Pisces. Both clusters have a redshift of 0.4, which corresponds to a look-back time of about 4,000 million years, if we assume for the moment that the Universe has an age of 15,000 million years. The images of the two clusters contain several hundred galaxy images, and were inspected by Alan Dressler and his co-workers. They discovered that the morphologies of these galaxies were identifiable with the Hubble classifications evident in galaxies in the present-day Universe. This was at once a significant and a surprising result, signifying that the galaxies which populated the Universe when it was two-thirds of its present age were highly developed objects. There were some differences, however, which were every bit as significant as the similarities. The cluster CL 0939 + 4713 contains many more spiral galaxies than do clusters of today (see Figure 8.10). Not only that, but the spirals are nearly all peculiar in some way

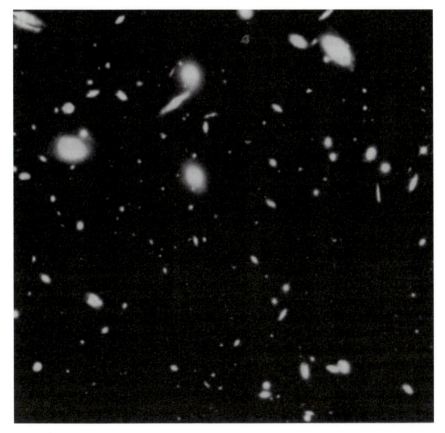

Figure 8.10. The distant galaxy cluster CL 0939 + 4713 is located in Ursa Major, at a distance of about five billion light years. It contains spiral galaxies which appear less distinct than those in the present-day Universe, but the elliptical galaxies are virtually identical. (Photograph reproduced courtesy of Alan Dressler, Carnegie Institution and NASA/STScI.)

or another. The question of what is producing the peculiarities remains slightly perplexing. At first, it seemed likely that the galaxies were still forming, but closer inspection of the images has led to the conclusion that the density of the cluster is probably pulling them apart. Certainly, some mechanism is responsible for the depletion of spiral galaxies in the dense clusters during our present epoch, and gravitational interaction would seem one of the best candidates. The only other possible mechanism is that hot gases in the cluster have stripped the star-forming gas from the spirals as they have travelled around the cluster.

The results also showed that the galaxies in Dressler's two distant clusters were much bluer than those of today, and star formation was a much more prevalent activity 4,000 million years ago. In the clusters examined, the blue galaxies were almost invariably the peculiar spiral galaxies. Richard Ellis and a team of UK astronomers used the HST to probe another cluster – this time at a redshift of 0.3 – and discovered that most of the blue galaxies were disturbed systems, either in the process of merging or showing the distinctive signs of interaction. On the whole, the Ellis cluster contained fewer spirals than did the first two. This suggests that, considering its greater age, it has had more time to dynamically evolve than have the first two.

More surprises were in store when the HST identified even more distant clusters. Mark Dickinson identified a cluster surrounding the radio galaxy 3C324 at a redshift of 1.2 (see Figure 8.11) – representative of the Universe when it was about one third

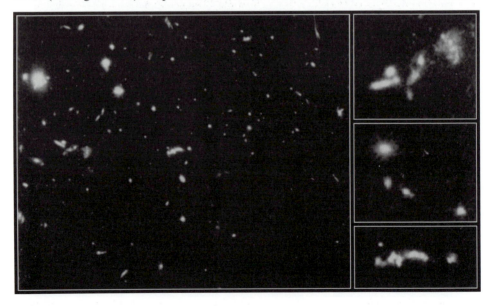

Figure 8.11. The cluster in Serpens, around the peculiar galaxy 3C324, is located at a distance of nine billion light years, and contains spiral galaxies which bear very little resemblance to those in the present-day Universe. However, the elliptical galaxies are remarkably similar. (Photograph reproduced courtesy of Mark Dickinson, STScI and NASA/STScI.)

Figure 8.12. One of the furthest galaxy clusters imaged so far by the Hubble Space Telescope is located in Sculptor and lies at a distance of about 12 billion light years. The individual galaxies in this cluster are very difficult to distinguish, but some of the visible fragments appear likely to become spiral galaxies. The elliptical galaxies are very similar to their present-day appearance. (Photograph reproduced courtesy of Duccio Macchetto, ESA/STScI, Mauro Giavalisco and NASA/STScI.)

of its present age. In this study, hardly any of the galaxies were recognisable as present-day spirals, although some resembled edge-on discs. Apart from an abundance of disturbed and merging galaxies, with fragments of galaxies strewn about the cluster, the biggest surprise was the discovery of mature elliptical galaxies. Analysis of these objects revealed that they were essentially the same as their counterparts in the modern Universe. They contained old, mature stars, and displayed no hint of ongoing star formation.

One of the most distant clusters was discovered by the Hubble Space Telescope. It is in Sculptor, and lies in front of the quasar Q0000-263 (see Figure 8.12). Located at a redshift of 3.3, the cluster is seen as it existed only 1,500–3,000 million years after the Big Bang. At this incredible distance, the distinction between the types of galaxies present in the cluster (about 14 individual members have so far been identified) become increasingly blurred. One galaxy, however, appears to display the characteristics of a mature elliptical – which is interesting. If this result is confirmed it will provide the tightest constraints yet on galaxy formation (see Figure 8.13). For the elliptical to be mature it must have condensed from the intergalactic medium – a

Figure 8.13. Comparison of different galaxies, such as those illustrated in Figures 8.10, 8.11 and 8.12, allows astronomers to compile an evolutionary sequence, in order to determine how galaxies evolve through time. (Photograph reproduced courtesy of A. Dressler, M Dickinson, D. Macchetto, M. Giavalisco and NASA/STScI.)

process which according to theory takes at least 1,000 million years. It then had to have time to generate its first population of stars, to generate any subsequent population before star-formation ceased, and then have time for its high-mass stars to explode and leave only the older red stars. The age of the cluster, as indicated by its redshift, whilst not impossible, leaves such a short time for the galaxy to evolve as we observe it (after first allowing approximately 300,000 years for the Universe to decouple) that some astronomers have become nervous! Others view it as indicating that the process of galaxy formation was not a slow one. As soon as gravitational collapse was possible, it began.

One of the most interesting objects in the image around Q0000-263 is a galaxy seen very close to the quasar. It is reminiscent of an object which may be the most distant nascent galaxy ever observed. There seems little doubt that astronomers have finally developed the technology to see the era of galaxy formation itself.

9

Deep fields and the formation of galaxies

The quest for protogalaxies is one of the hardest tasks in modern astronomy. A protogalaxy has, at best, blurry definition. In stellar astrophysics a protostar can be defined as any young stellar object which has yet to begin fusing hydrogen into helium at its centre. But when does a protogalaxy become a galaxy? When it begins to shine with starlight perhaps? What then distinguishes a cloud of hydrogen gas from a protogalaxy? Despite these vagaries of terminology, one thing is certain: protogalaxies must be faint. Combined with their expected great distances, this makes them very difficult to observe.

9.1   PROTOGALAXY CANDIDATES

Observations of the quasar BR 1202-07, by a team of astronomers led by Sandro D'Odorico, revealed an interesting feature: its spectrum included an absorption line at 654.5 nm, the properties of which suggested that there is a massive cloud of hydrogen at a redshift of 4.38. The quasar itself exists at a redshift of 4.69. This feature appeared to be an incredibly distant Lyman-α system. Subsequent spectrographic observations of the quasar revealed that light was being blocked not only by hydrogen, but also by other elements. Metal absorption was found in the spectrum at the same redshift as the hydrogen. This immediately suggested that the hydrogen was not simply an inert cloud of gas that was just beginning to show the effects of a 'first round' of star formation. Perhaps it really is a protogalaxy. Analysis of the strengths of the spectral lines, by Limin Lu and co-workers, of Caltech – has shown that the chemical content of the protogalaxy indicates that stars have been forming for only a few tens of millions of years – a baby, in galaxy terms!

The optical identification of this object, however, is difficult. Following these tantalising glimpses in the spectrum of the quasar, astronomers wanted to obtain an actual image of this protogalaxy. The absorption lines revealed its silhouette; and now astronomers wanted to look it straight in the face.

An image obtained with the Keck 10-m telescope on Mauna Kea did indeed reveal a faint, diffuse object, two arcseconds from BR 1202-07. It looks superficially similar to the young galaxy near the quasar Q0000-263 in Sculptor. An initial analysis showed it to possess the correct wavelength dependence, but successive observations have cast doubt on its identity with the absorbing protogalaxy. An infrared image taken with the University of Hawaii's 2.2-m telescope shows it to possess the same redshift as BR 1202-07. A Hubble Space Telescope image of the objects suggested that the 'protogalaxy' was actually nothing more than a cloud of gas, sub-galactic in size, illuminated by the nearby galaxy. So where is the true protogalaxy? It must be there somewhere because of the absorption lines in the quasar's spectrum. Astronomers must simply look harder.

Searches for protogalaxies can also take the form of searching at specific wavelengths. For example: a protogalaxy which is forming its first generation of stars must be heavily laden with hydrogen gas. The supermassive stars would ionise the gas and make it glow, and the galaxy would take on the appearance of a massive emission nebula. If this were to occur in an object at the distance of the quasars, the light from the ionised gas would be redshifted into the near-infrared portion of the electromagnetic spectrum. Using this rationale, Matthew A. Malkan and colleagues identified a potential protogalaxy on the Sextans–Leo border. It is at a redshift of 2.5, so is not as far away and difficult to study as some of the other protogalactic candidates. It displays the hallmarks of a young galaxy system by having excessive hydrogen and nitrogen emissions. Astronomers are anxiously awaiting its more detailed spectroscopic study.

9.2   FAINT BLUE GALAXIES

Whilst galaxy formation is traditionally thought to have taken place in distant realms, some astronomers have turned their attention to slightly closer regions in the hopes that the answer may also be found there. One reason for this is that the advent of sensitive detectors and telescopes has not only allowed us to probe further into the Universe, but has also provided us with the ability to carry out much more complete surveys of our more immediate surroundings. The Medium Deep Survey on the Hubble Space Telescope uses the Wide Field/Planetary Camera 2 to take a photograph of the sky whenever one of the HST's other instruments is in use. In this way, a random survey of the sky is possible, and does not interfere with the HST's more specific observations. Tens of thousands of galaxies have been identified on 50 random snapshots. Of these, a very large number are faint, blue and irregular in shape (see Figure 9.1). Follow-up observations to image these objects in detail showed that faint blue galaxies (as they were termed) are probably the most ubiquitous type of galaxy in the Universe. There are so many of them, in fact, that if our eyes were sensitive enough we would be able to detect a faint, blue background glow across the whole night sky.

The faint blue galaxies seem to predominate at distances of 3,000–8,000 million light years, but are incredibly scarce in the Universe today. Obviously the star formation which distinguished them in their early existence has been extinguished or, perhaps, the galaxies themselves have been subsumed or dissipated by interactions with others.

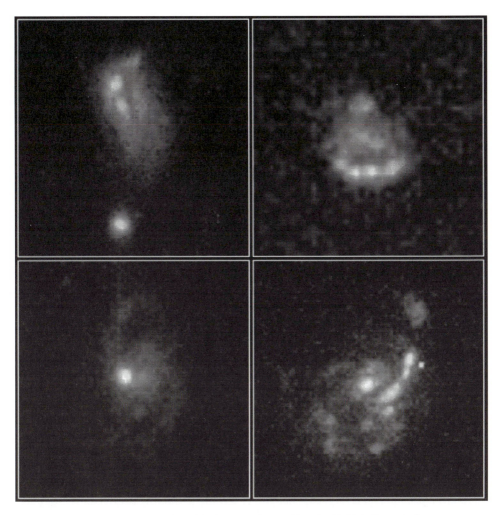

Figure 9.1. These peculiar blue galaxies were imaged by the Hubble Space Telescope as part of a three-year Medium Deep Survey. They are so distant that they are being seen as they were when the Universe was about half its present age. The Medium Deep Survey has revealed a bizarre variety of structures in these distant galaxies, which are irregular in shape and remarkably blue in colour. The blue colour indicates that vigorous star formation has recently taken place in these objects, which once far outnumbered galaxies like our own Milky Way but have since faded. (Photograph reproduced courtesy of Richard Griffiths, the Medium Deep Survey Team and NASA/ STScI.)

Another form of galaxy which has come to the forefront of attention in the last ten years is the low surface brightness galaxies (LSBs). A few far-sighted astronomers have cautioned against assuming that galaxies only take the forms seen in the present-day Universe. For instance, in the 1930s Fritz Zwicky suggested

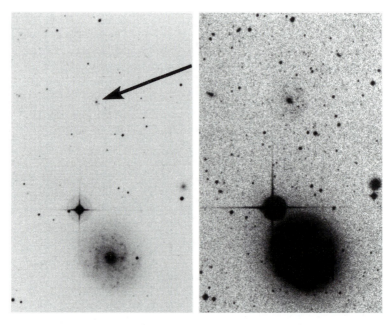

Figure 9.2. The discovery plates of the low surface brightness galaxy Malin 1 (arrowed). The image at left is the unprocessed plate, whilst the image at right has been contrast-enhanced to show that Malin 1 is a low surface brightness galaxy. (Photographs by David Malin, from plates taken with the UK Schmidt telescope.)

that the easily observable galaxies may be only a sub-set of the total population. Halton Arp also made a similar comment in the 1960s, when he asserted that the galaxies which were known to exist actually displayed a very narrow range of properties and that, perhaps, others with a much wider parameter base were out there, awaiting discovery. In 1986, world-renowned astrophotographer David Malin noticed that as he used photographic amplification methods on his image plates he could detect faint galaxies 'everywhere'. Analysis of one of these 'blobs' by Chris Impey and Greg Bothun showed that it was about 700 million light years away and about six times the diameter of the Milky Way! Its mass was about 20 times that of the Milky Way, and unquestionably Malin 1 (as it was called) was the largest spiral galaxy so far discovered in the Universe (see Figure 9.2). But why then was it so faint? It just did not seem to want to form stars. It contained so few stars that, if it were displaced to the position of the Andromeda galaxy – at a distance of 2.2 million light years – it would be the width of 40 full Moons; yet we would hardly notice it was there. Its lack of stars is so pronounced that it would glow only at a feeble 2% brighter than the sky.

A survey undertaken by Impey and co-workers in 1991 revealed 500 previously unknown low surface brightness galaxies in an equatorial strip of the sky. All the galaxies discovered were within the distance range of 200–400 million light years. There are so many, in fact, that if this trend continues throughout the Universe, the galactic content of the cosmos may have been underestimated by 100%!

Radio observations of the low surface brightness galaxy UGC 6614 show conclusively that it contains a vast amount of gas, but for some reason it is not condensing into stars. Theoretical calculations have shown that a low-density gas disc may be stable against the kind of fluctuations which lead to gravitational collapse and star formation. Certainly, as far as galaxy formation goes, the denser the initial cloud the faster the collapse will proceed; so perhaps low surface brightness galaxies are simply galaxies which have taken much longer to form than the others because they originated from lower-density gas clouds. If so, they offer the unique opportunity to study a galaxy in the process of formation.

If LSB galaxies are indeed formed from small density fluctuations rather than from the large fluctuations which led to luminous galaxies, they are actually much better tracers of mass in the Universe. To use an analogy: luminous galaxies are the white-water swells on top of the greatest waves, whilst the LSB galaxies trace out the smaller-scale swells which trace, much more effectively, the distribution of water.

Another idea about these objects links them with the faint blue galaxies. Some think that perhaps low surface brightness galaxies are all that remains of the faint blue galaxies once that initial burst of star formation has run its course. If this is proved to be the case, then astronomer Ed Saltpeter would like to call them 'grinning Cheshire cat' galaxies! He considers that the low surface brightness galaxies are the cosmic equivalent of *Alice in Wonderland*'s Cheshire cat because, despite being bright to begin with, they are now fading to obscurity, and all that remains is the galactic equivalent of the Cheshire cat's grin!

9.3   DEEP FIELDS

Our deepest views of the Universe have been supplied by the Hubble Space Telescope. The first of them – the Hubble Deep Field (HDF) – was obtained by the orbiting telescope between 18 and 28 December 1995, as it peered at a tiny 'keyhole' area of the Universe just above the asterism of the Plough (the Big Dipper). It is a composite image of 342 separate exposures, taken at four different wavelengths, of this same field of view. It has been image-processed to reveal incredibly faint galaxies of about magnitude 30 (6,000 million times fainter than humans can see with the naked eye). The image contains roughly 3,000 galaxies, all at various stages of their evolution (See Figure 9.3).

One of the first tasks was to establish which galaxies in the image are nearby and which are distant. Traditionally this would have been done by measuring the redshift of the galaxy (as described in Chapters 4 and 6) using a spectroscope. Taking a spectrum of a galaxy spreads its light over a greater area on the CCD detector, however, and so renders it even fainter at each wavelength. The Keck 10-m telescope can only perform spectroscopy on galaxies that are around magnitude 24; the faintest galaxies on the HDF image are therefore 100 times fainter than objects from which we can obtain spectra, using current technology. It is therefore not possible to obtain spectra for all of these newly discovered galaxies, and astronomers are having to rely on the photometric data (responsible for producing the colour image) to assist them.

Figure 9.3. The Hubble Deep Field (HDF) will go down in history as one of the most important images in modern cosmology. Although the entire image covers only a tiny area of the sky 1/140th the diameter of the full Moon (about 25% of the HDF is shown here), it represents a narrow 'keyhole' view all the way to the visible horizon of the Universe. It is a snapshot of the Universe, extending from the present day to almost the point of its creation. It contains images of approximately 3,000 previously unseen galaxies. (Photograph reproduced courtesy of R. Williams and the HDF Team (STScI) and NASA.)

To carry out this analysis, astronomers exploit features in the hydrogen atom's spectrum known as ionisation edges. Figure 2.8 showed schematically the effects of ionisation edges on a black-body spectrum. The loss of an electron from the hydrogen atom is caused when a photon of sufficient energy is absorbed by the atom. Thus it is expected that, if a cloud of hydrogen gas is illuminated by a continuous spectrum, the photons which contain sufficient energy to ionise the hydrogen will be absorbed, and those which do not contain enough energy will not be absorbed. Photons which contain rather more energy than necessary to ionise the hydrogen will still be absorbed, as the electron will simply convert the excess energy into kinetic energy. Thus, an observation of light that has passed through the hydrogen cloud will show sudden drops in intensity across these ionisation edges. These are sometimes referred to as 'breaks' in the spectrum.

Hydrogen has two such useful breaks in its spectrum, created by the ionisation edges. One is located at 400 nm, and is caused by the ionisation of electrons that are in the first excited state. The second is located at 91.2 nm, and is caused by the ionisation of electrons that are in the ground state. Because hydrogen is the most abundant atom in the Universe, it is expected that galaxies – being made predominantly of hydrogen – will display very obvious hydrogen ionisation edges. If the galaxy in question is far away, then the cosmological redshift will have altered the wavelength at which the ionisation edges are observed to occur, during the light's long passage to Earth.

The HDF was obtained at four different wavelengths: ultraviolet (300 nm), blue (400 nm), yellow (550 nm) and red (800 nm). The ionisation edges will therefore cause the galaxy to appear significantly less bright with some filters than with others. In extreme cases the galaxy appears to simply disappear (see Figure 9.4).

In Chapter 5 we introduced the notion of the scale factor of the Universe, $R(t)$, which is an arbitrary measure of the size of the Universe, and changes with time, t. If we take a known wavelength, λ, to be our scale factor, the ratio of this to its currently observed wavelength, λ_0, produces the factor by which the Universe has expanded since the time of the radiation's release. The redshift equation (4.6) also contains this ratio, and we can therefore easily equate the redshift, z, with the expansion factor of the Universe:

$$1 + z = \frac{\lambda_0}{\lambda} \tag{9.1}$$

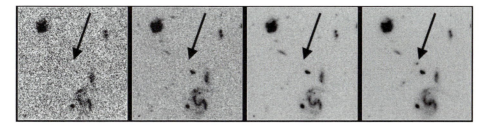

Figure 9.4. Photometric redshifts obtained with different coloured filters and centred on different wavelengths. The arrowed galaxy is visible only on the red plate, indicating that the 91.2-nm ionisation edge has been redshifted by a factor of approximately 5. Galaxies visible through all filters are at much smaller redshifts. (Photograph reproduced courtesy K. Lanzetta and A. Yahil (SUNY) and NASA.)

The quantity $1 + z$ is the spectral ratio, which translates directly into the expansion factor of the Universe. In other words, radiation from objects which possess a cosmological redshift of 1 originated at a time when the Universe was half its present size, because the spectral ratio is 2. It also means that the radiation released by the galaxy will have been doubled in wavelength by the cosmological redshift. A galaxy at redshift 4 will therefore have had its radiation lengthened by a factor of 5.

Hence it becomes obvious that on the HDF the filter which does not reveal the galaxy depends upon the redshift of the galaxy. The break at 400 nm is useful only out to a redshift of 1, because the longest filter wavelength on the HDF is 800 nm. Luckily there is the other spectral break – the Lyman break, at 91.2 nm. This must be lengthened by spectral ratios of 3 and 8 (redshifts of 2–7) in order for it to be detectable on the HDF. Redshifts, and therefore distances, can be estimated for all galaxies on the HDF – an approach referred to as the photometric redshift method.

A team of astronomers led by Kenneth Lanzetta has found six galaxies which are present on the red plates but which fail to appear on the shorter wavelength images. This is a strong clue that they are distant objects which have been severely redshifted. They appear to lie at redshifts greater than 5, which exceeds the record set by the most distant known quasar.

It is not just distances that can be gauged by the colours of these galaxies; the rate of star formation can also be estimated. When stars form, a small number of high-mass stars form together with a large number of low-mass stars. The high-mass stars emit ultraviolet and blue light, and the low mass stars emit yellow and red light. By estimating the redshifts to the HDF galaxies, astronomers are able to determine which wavelengths on the images represent light that was originally emitted at ultraviolet wavelengths. Therefore, even if the light being observed has been cosmologically redshifted to visible yellow wavelengths, it is known that it represents ultraviolet emission (restframe ultraviolet) from the galaxy.

By studying the amount of restframe ultraviolet emitted by a galaxy, the number of high-mass stars that the galaxy contains can be estimated. These stars are special, because they live for only a few tens of millions of years, and therefore offer a reasonably good yardstick with which to measure the rate of star formation in the galaxy.

By separating the galaxies into redshift ranges and then analysing them, the average rate of star formation within each redshift range can be estimated. This type of work was first undertaken by using ground-based telescopes to observe galaxies that are much closer than those on the HDF. The survey, headed by Professor Simon Lilly, of Toronto University, is known as the Canada–France Redshift Survey (CFRS), and has reached galaxies out to a redshift of about 1. Most of the galaxies on the HDF are estimated to be at higher redshifts of 3–5. Together, the CFRS and the HDF data provide the first indications of the star-formation history of the Universe (see Figure. 9.5). The rate of star formation must have peaked at about redshift 2, corresponding to an age of the Universe of 2–3 billion years. As for the actual rate itself: the data suggests that 10–15 times more stars were forming then than are forming today.

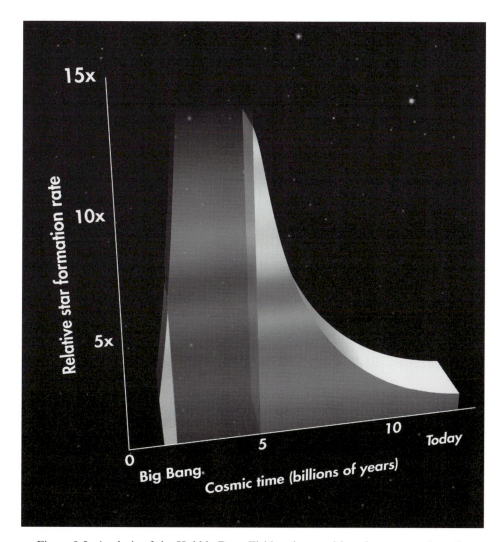

Figure 9.5. Analysis of the Hubble Deep Field and ground-based surveys such as the Canada–France Redshift Survey has led to the conclusion that star formation in the Universe peaked billions of years ago. The exact rate of star formation is unknown at present, but is thought to be 10–15 times its present value. (Photograph reproduced courtesy of P. Madau (Space Telescope Science Institute), J. Gitlin (Space Telescope Science Institute) and NASA.)

The area of sky covered by the HDF was minuscule – only 1/140 the area of the full Moon – and astronomers could therefore not be certain that what they were seeing on that one image was representative of the Universe as a whole. So, in order to increase their confidence, a second deep field image was obtained by pointing the telescope in the opposite direction from the Plough (Big Dipper) towards the southern constellation of Tucana.

Figure 9.7. The Very Large Telescope, in Chile – seen here whilst still under construction – is the spearhead of a new generation of high-class telescopes that will probe the far-distant realms of the Universe. (Photograph reproduced courtesy of ESO.)

The Hubble Deep Field South (HDFS) was obtained in October 1998 (see Figure 9.6, colour section). For two weeks the HST observed a region of sky in Tucana. It took 995 separate images of the location, which were then combined into the final picture. Each separate observation had an exposure time of between 30 and 45 minutes. Observations were made at wavelengths in the ultraviolet, optical and infrared regions of the spectrum, and the HST discovered more than 2,500 new galaxies in this tiny portion of sky. This time the deep field location also contained a quasar, so the Lyman-α forest could be particularly well studied. The actual extraction of science from the images will take months, if not years, and will rely on follow-up observations from southern hemisphere telescopes such as the Very Large Telescope (VLT), built by the European Southern Observatory (ESO) and located in Chile (see Figure 9.7).

Perhaps the most obvious and best feature about the HDFS is that it appears to be very similar to the original Hubble Deep Field. This is good news, because it supports the idea that the Universe is indeed isotropic and homogeneous (as was discussed in Chapter 4). Astronomers can therefore be more confident about drawing conclusions about the whole Universe, based on these two deep fields. This is very encouraging, because to survey the whole sky to a similar depth (a limiting magnitude of 30) would take the HST 900,000 years!

As with the original deep field, astronomers will be most anxious to determine the distances of the galaxies, and then calculate the star-formation rate and the change in the shapes of galaxies with time.

With the bulk of evidence now supporting the validity of the conclusions from the original deep field, an interesting problem has arisen. Studies of elliptical galaxies in the present-day Universe reveal that these gigantic stellar collections are packed full of older stars, which must have formed very early in the history of the Universe. Even though deep fields show a pronounced rise in the rate of star formation, however, it does not appear to be enough to explain why there are so many 'old' stars in the Universe today.

A particular study of the Hubble Deep Field was undertaken by Michael Vogeley, of Princeton University Observatory, and his team of collaborators. They were examining the image to determine whether even fainter (and so even more distant) star-forming galaxies were visible on the HDF. Their method involved digitally removing all the known galaxies from the image so that they simply had the background space to work with. If fainter galaxies – just at the limit of detection – were present, they would cause the background sky to appear slightly brighter close to them, and dimmer where there were none. Vogeley's analysis showed no such variations in the sky brightness of the image, however, confirming that there are no other major sources of optical radiation in the Universe. So, where are the rest of the star-forming galaxies?

9.4   THE LONG-WAVELENGTH UNIVERSE OF GALAXIES

The further one looks into the Universe, the more one suffers from the redshift. So, the next step in the 'deep field' business is to begin observing at infrared wavelengths. This allows the restframe ultraviolet and optical emission from even more distant objects to be seen more easily. For example: if a galaxy is located at a redshift of 5, the wavelength of peak light emission from a Sun-like star (550 nm) will be stretched into the infrared (3.3 µm). Another reason for moving to even longer infrared wavelengths is that stars are thought to form in very dusty environments, and the dust emits infrared wavelengths of radiation. Obviously, these too will be redshifted – depending upon how far away the galaxy is located – but the dust emission is more likely to remain within the infrared region of the spectrum, because the region is so broad. Typically, the infrared emission by dust begins at wavelengths of about 5 µm, and can still be readily detected at wavelengths of a few hundred µm – all still within the infrared range. Perhaps the majority of stars which have formed in the Universe formed very early inside very dusty protogalaxies, which could be the reason why they have not been detected by optical surveys such as the deep fields.

According to this theory, galaxies beyond redshift 5 would be almost invisible at optical wavelengths because they have not yet formed enough stars to blow away their dusty cocoons and render them fully visible. These galaxies should, however, be superb emitters of infrared radiation, as they will be full of young, forming stars that are enshrouded in dust that will be absorbing their radiation and re-emitting it copiously in the infrared. So, a sensitive infrared camera should be able to see beyond redshift 5 into the realm where galaxies first formed.

In the next chapter, the COBE satellite will be described. One of its experiments has direct relevance here, however. The Diffuse Infrared Background Experiment (DIRBE) was designed to search for the faint infrared glow produced by the formation of galaxies. DIRBE scanned half of the sky every week for a ten-month period between December 1989 and September 1990. It surveyed the sky at ten infrared wavelengths from 1 μm to 240 μm. After the data were combined to produce ten full-sky maps (one at each of the different wavelengths), the infrared emission from dust in our Solar System, and the infrared radiation from the Galaxy, was removed from the images. What remained was analysed, and on the 120 μm and 240 μm images there was discovered to be a faint excess of infrared radiation (see Figure 9.8, colour section). The excess was most easily seen in 'windows' near the north and south poles of the Milky Way. Through windows, astronomers have a relatively unobstructed view across billions of light years.

The discovery of this background glow of infrared – thought to be coming from dust-enshrouded protogalaxies – has prompted great interest in the observational study of this region of space. As a first step towards the study of the Universe beyond redshift 5, a team of astronomers using data collected by ESA's Infrared Space Observatory found 24 candidates for galaxies that would be contributing to this infrared background radiation.

This type of study was to have been extended by the Wide Field Infrared Explorer (WIRE), which was to be placed in orbit during March 1999. It was intended to survey the sky at two infrared wavelengths – 12 μm and 25 μm – to search for starburst galaxies. This would have allowed it to discover just how much infrared energy is being emitted in total by the starburst galaxies, the speed at which the starburst galaxies are evolving and how common protogalaxies are at redshifts of less than 3. Essentially, WIRE would have attempted to discover whether galaxy formation is an ongoing process rather than a process that is confined to the very far-distant Universe.

Unfortunately, shortly after launch WIRE went it an uncontrolled spin. Engineers on the ground worked round the clock to solve the problem, but it soon became obvious that the spin was being produced by hydrogen gas leaking from the satellite. This hydrogen was the precious coolant that would be used to chill the detectors to their operational temperature. For days the satellite continued to spin, and eventually all of the hydrogen was vented into space and the instrument was essentially useless. The work will now have to be carried out by new satellites – such as NASA's Space Infrared Telescope Facility (SIRTF) – which are currently being built (see Figure 9.9).

From the ground, the longest wavelengths of infrared (those that 'nudge' the radio spectrum) can be observed. The James Clerk Maxwell Telescope on Mauna Kea was recently the focus of much attention when a new detector called SCUBA (Submillimetre Common User Bolometer Array) was commissioned. Using this instrument, two teams of astronomers – one from Britain and the other from Japan and America – discovered a hitherto unknown population of star-forming galaxies at extreme distances. This new galaxy population contains very dusty objects, and it is the dust which obscures their star formation. In this respect, they are probably typical protogalaxies. The radiation from the stars heats up the dust grains, and they

Figure 9.9. The Space Infrared Telescope Facility (SIRTF) – currently being constructed by NASA – is a satellite designed to study the Universe at infrared wavelengths. This is a simulated image of what it would probably see if it were to observe the same location of space captured in the Hubble Deep Field. Because it is taken at infrared wavelengths of 3.5–8 μm, the emission is less affected by interstellar extinction than are the optical images. This enables more accurate determination of the galaxies' ages and masses. (Photograph reproduced courtesy of NASA.)

re-radiate this energy, but at far-infrared wavelengths. Because the galaxies are so distant, the cosmological redshift stretches the already long infrared wavelengths to almost a millimetre.

The radiation from these submillimetre sources implies that some of the objects are forming stars at a rate 10–100 times greater than are galaxies at present. This is a

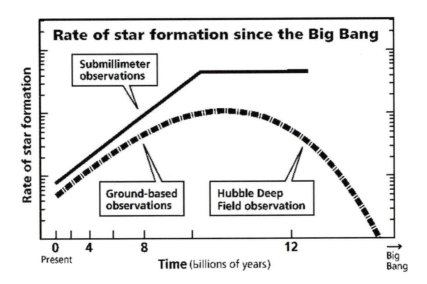

Figure 9.10. New observations of the very distant Universe, taken at millimetre wavelengths, suggest that there is an entire population of young, distant galaxies that astronomers have not yet detected. They are very dusty objects, and the new stars are enshrouded by material. The formation is detectable only after their light has been absorbed and re-emitted as long-wavelength infrared radiation. It is detected in the millimetre wavebands because of the cosmological redshift. This discovery suggests that the star formation rate did not rise suddenly and peak at redshifts of around 2, but that it was steady from a much earlier epoch in cosmic history. (Diagram reproduced courtesy of NASA/STScI.)

phenomenal rate, and is certainly sufficient to account for the stars in elliptical galaxies today (see Figure 9.10).

9.5   THE DARK AGES

The bulk of the available evidence points to galaxies and galaxy clusters forming before the Universe reached an age of 2,000–3,000 million years. In terms of redshifts, the most active period of galaxy and cluster formation therefore occurred between redshifts of 5 and 1,000. This period is called the dark ages, because we have no direct observational data from this epoch of universal history. In this realm we have to rely almost solely on theory. Remarkably, we can make progress and constrain our ideas about galaxy formation by simply considering a few elementary facts about the Universe around us.

If we accept the common-sense assumption that, in order for a galaxy to form, there must be enough space to accommodate it, then we can set a maximum redshift at which galaxy formation can begin. The average diameter of a galaxy is about 20 kpc, and so for a galaxy to be able to form, there must have been at least 20 kpc between the protogalaxies. At lesser distances than this, the galaxies can no longer retain their identity as separate objects. If we assume an average distance of 1 Mpc between galaxies in the present Universe, then we can use this to represent our scale factor measurement. This implies that individual galaxies could not have formed before redshift 50. Larger structures such as clusters would have to emerge even later. This does not imply that the fluctuations in density – which eventually become individual galaxies and clusters – were not present at earlier times in the Universe. The assertion is that the individual galaxies cannot have emerged in their present form until the Universe was large enough to accommodate them.

This assertion forces us to consider the morphology of the fluctuations which grew into galaxies and larger structures. For the moment we will not concern ourselves with the physical mechanisms responsible for the creation of the fluctuations; instead we shall simply accept that they exist, and chart their progress in the decoupled Universe.

We can be reasonably certain – from both theory and observation – that the collapse of galactic structures occurred on all levels simultaneously. Thus, as stars are forming, so galaxies are condensing and clusters are aggregating. There are, however, subtleties (discussed in the next chapter) which lead to two different formation scenarios, but broadly speaking the evolution of universal structure proceeds on many levels. As we have seen throughout this book, the Universe appears to have begun in a nearly homogeneous state. Fluctuations from homogeneity can be charted by considering how the density, ρ, at each point in the Universe, x, deviates from the average density of the Universe, ρ_0:

$$\delta\rho(x) = \rho(x) - \rho_0 \tag{9.2}$$

A dimensionless quantity known as the density contrast, $\delta(x)$, can then be defined and used to trace the growth of the fluctuations.

$$\delta(x) = \frac{\delta\rho(x)}{\rho_0} \tag{9.3}$$

Whilst the collapse proceeds as a scale-independent process, the growth is linear. This is a very slow process at first, for various reasons. Firstly, the initial fluctuations are very small (as will be seen in the next chapter). Secondly, the expansion of the Universe acts to inhibit the condensation of the fluctuation. Condensation will also be inhibited by the internal pressure of the matter, which acts like a gas, in the fluctuation. If it were not for the expansion of the Universe, the growth would proceed at exponential rates and there would be no linear growth phase.

To determine the mass limit – which defines whether or not a fluctuation will collapse gravitationally – the kinetic energy of the gas present in the fluctuation must be considered. This kinetic energy is produced by the expansion of the Universe and the internal gas pressure, so by equating this with the gravitational potential of the fluctuation, the criterion for collapse can be quantified. If the kinetic energy is greater than the gravitational potential, then the fluctuation will not grow. Instead it will dissipate by behaving like a sound wave in the dense early Universe. The borderline is defined when both kinetic and potential energies are equal.

$$\frac{GM^2}{R} = \frac{MV^2}{2} \tag{9.4}$$

where $G = 6.672 \times 10^{-11}$ Nm^2/kg, M is the mass of the volume element under consideration, V is a measure of the random motions in the gas which are resisting the collapse, and R is the radius of the fluctuation. By treating the problem hydrodynamically, the mass at which collapse is unavoidable can be derived. This is the Jeans mass, M$_J$, named after the distinguished astrophysicist James Jeans, who first derived it.

$$M_J = \frac{4}{3} \pi \left(\sqrt{\frac{\pi V^2}{G\rho}} \right)^3 \rho \tag{9.5}$$

This equation also includes a density term, ρ. As soon as this mass is attained by a fluctuation, the region is destined to experience gravitational collapse. In the linear phase of its growth, a fluctuation still does not contain the mass necessary to prevent it from experiencing the expansion of the Universe. If one imagines that the rate at which the Universe expands is governed by the density of space at each point throughout its volume, then $\delta(x)$ can be used to determine the local velocity of expansion. The greater $\delta(x)$, the more work the Universe has to do to overcome the gravitational potential at that point before it can expand. Therefore, in the region of fluctuations, the universal expansion is slower than in less dense areas.

Whilst in the linear growth regime, this slowing down is at first almost imperceptible – so much so, that the diameter of the co-moving volume element, l_{co}, which contains the fluctuation, continues to grow at the rate of universal expansion. Thus, its linear diameter at any redshift epoch can be expressed as

$$\frac{l_{co}}{l} = 1 + z \tag{9.6}$$

As the fluctuation approaches the non-linear growth phase, it finally accumulates enough mass to be able to resist the expansion of space. At this point, the material contained within the fluctuation begins to condense. The point at which the fluctuation attains enough gravitational potential to resist the universal expansion and begin collapsing is known as the point of turnaround. This criterion for gravitational collapse uses mass as its defining quantity and, because it is impractical to deal with all scales of fluctuations at once, they are usually also divided up by mass. 10^6 solar masses defines a fluctuation which will become a dwarf galaxy/ globular star cluster, 10^{11} solar masses defines a galaxy, 10^{12} solar masses defines a globular cluster, and 10^{14} solar masses defines a supercluster.

Different fluctuation scales reach their non-linear growth phase at different points in the history of the Universe. In the absence of real data, these turning points are defined by arbitrarily setting $\delta(x) = 1$. The point at which the growth of most fluctuations is thought to have become non-linear occurred somewhere between redshift of 100 and 50; although, in the present-day Universe, there is evidence that galactic structures on scales of greater than 8 Mpc are still in a linear growth phase. The formation of large-scale universal structure may therefore not yet be complete.

As soon as the non-linear phase is entered, the objects begin the headlong collapse which leads to the formation of protogalaxies and other objects (discussed in the earlier sections of this chapter). The details of galaxy formation are, however, still very unclear. The first important question we have to answer is: what is the order of formation? Did the galaxies form first and then aggregate into clusters, or did the various large-scale structures emerge first and later fragment into individual galaxies?

The origin of the primordial fluctuations must also be addressed. Where did they come from, how were they formed, and what was their behaviour during the very early Universe before the decoupling of matter and energy? In order to answer these questions, the only observational evidence available to guide us comes from the very edge of the observable Universe itself. The answers to these tantalising questions may be locked up in the cosmic microwave background radiation.

10

The glow from the edge

Although the actual formation of galaxies is, tantalisingly, just beyond our observational reach, the fluctuations which eventually became galaxies and clusters are available to us. This information is accessible to us due to our studies of the cosmic microwave background radiation. The photons which compose this radiation provide us with a view, as far back as we can possibly observe, using conventional technology. Our journey has finally brought us to the faint glow which marks the edge of the Universe.

The redshift is approximately 1,000 from our point of view and shows us the Universe just 300,000 years after the Big Bang. The maelstrom from which the radiation was set free had cooled to about 3,000 K – almost cold, considering that the temperature of the Universe, at less than a second in age, had been many millions of Kelvin. As we observe it now, it is a frigid 2.7 degrees above absolute zero.

Chapter 4 described the appearance of the cosmic microwave background as a thermal source of radiation. Having established this as an observational fact, and therefore using it as a constraint in our model of the Big Bang, it described the evolution of the Universe up to the point at which the background was released. We shall now concentrate on the ways in which the background's spectrum deviates from its black-body characteristics, and discuss the consequences of these deviations in our larger picture of how structure emerged in the Universe.

10.1 THE COSMIC MICROWAVE BACKGROUND DIPOLE

There are many potential sources of anisotropy in the cosmic microwave background radiation (CMBR). If the Universe is not subject to a largely homogeneous expansion, then the variation in the Hubble constant along different lines of sight will correspond to differences in the temperature of the microwave background. In this non-isotropic view of the Universe, less dense lines of sight will expand faster, and the CMBR will appear more redshifted, so that it appears cooler in those lines of sight. Overly dense regions will expand only slowly, and therefore will appear hotter because they are not so redshifted. Although the equations of general relativity allow all manner of anisotropic expansion models, by the cosmological principle we assume that the Universe is isotropic, and so the CMBR is not subject, in any significant manner, to this form of fluctuation. As explained in Chapter 3, whilst there are local deviations from isotropy these are expected to average on the scale of the Universe as a whole.

If the Universe contains large-scale magnetic fields – such as those postulated by the plasma cosmologies discussed in Chapter 6 – then these will also alter the background radiation. As we are considering only Big Bang cosmologies in detail in this book, however, our discussions of magnetically induced anisotropies will proceed no further.

The two forms of anisotropy that we will consider in detail are those which are produced by the movement of our observing platform and those which are produced by inhomogeneities in the matter content of the Universe. The simple act of observation has shown that inhomogeneities exist on many cosmic scales – from galaxies to clusters, and from superclusters to voids. We would expect the concentrations and rarefactions of matter from which these structures sprang to have left an imprint somewhere on the microwave background.

Anisotropies in the cosmic microwave background are quantified by the ratio of their deviation, ΔT, to the average temperature of the radiation, T:

$$\frac{\Delta T}{T} \tag{10.1}$$

The most obvious anisotropy revealed in the microwave background is a dipole anisotropy which occurs at a $\Delta T/T = 0.005$ (see Figure 10.1, colour section). Assuming that this dipole is caused by our movement relative to the CMBR, then the isotropy is caused by the Doppler effect. We would appear to be moving at 360 ± 20 km/s in the direction of the border between Leo and Crater. This is a slightly puzzling result, because observations have clearly shown that the Sun is travelling around the centre of the Milky Way with a velocity of about 240 km/s, but in almost the opposite direction from Leo! Thus, it must be the movement of the centre of the Galaxy which is causing the anisotropy in the CMBR. Although the velocities are high, they are not yet relativistic and so we can simply add them, to discover that the Galaxy must be moving at roughly 600 km/s. Rigorous calculations – which take into account the Sun's movement around the Galaxy and the Galaxy's movement around the centre of mass of the Local Group – have confirmed this velocity. Based upon the interpretation of the microwave background's dipolar anisotropy, astronomers can say with confidence that, based upon observations from Earth, the Local Group of galaxies is moving with a velocity of 600 ± 30 km/s towards the Crater/Hydra border at Right Ascension 10.5 hours.

The source of this velocity has important cosmological consequences. If it were produced by primordial velocity fluctuations of the matter in the early Universe, then those velocities must have been much larger in the past, with reference to the co-moving frame of reference, which describes the expansion of spacetime. They will in fact be scaled in accordance with the spectral ratio, $1 + z$, and so at the point of the decoupling (z = 1000) would have been approaching the speed of light! There is absolutely no observational evidence for these kinds of motion in the Universe at decoupling. The small-scale anisotropies produced from regions of space which could be travelling at close to the speed of light would be enormous. The anisotropies that have been discovered (as we shall discuss later) are far too small in amplitude to have been produced by relativistic emitting regions.

The only other explanation for this motion is that the Local Group has been accelerated since the decoupling. The only physical mechanism which could possibly move a cluster of galaxies is that of gravity. For this to occur, the Universe has to be inhomogeneous on relatively large scales. Chapter 4 discussed this point, and concluded that inhomogeneities exist on scales of up to and more than 100 million light years.

In Chapter 8 the concept of streaming motions was introduced. These are large-scale, bulky flows of material (typically galaxies and the intergalactic medium) which are in motion because of the gravitational attraction of a nearby concentration of mass. In our local galactic neighbourhood, an enormous concentration of matter has been named the Great Attractor because of the effect it is having on the rate of the Hubble expansion in our local region of space. When the Hubble flow is subtracted from the individual galaxies' motions it can be seen that galaxies are streaming towards the Great Attractor.

Although this approach is promising, there is as yet no agreement between the velocity derived from the CMBR measurement and the direction of the Great Attractor, calculated from streaming motion studies. It is important to remember that the CMBR dipole measures our velocity relative to very distant matter which was present at the decoupling of matter and energy. The Great Attractor, by comparison, studies the distribution of relatively close-by material.

The presence of these inhomogeneities (as proved by the CMBR dipole) indicates that at some level there should exist the fossil imprints of the density differences on the microwave background. Small fluctuations which then grew in size to become large fluctuations are much more easily explained than fluctuations which were originally large.

10.2   RIPPLES IN THE COSMIC MICROWAVE BACKGROUND

The breakthrough in the search for the perturbations which were the seeds of today's structure came in 1992. NASA's satellite, the Cosmic Background Explorer (COBE) (see Figure 10.2) had collected data which, when processed, produced the first tantalising hints that the background radiation is anisotropic on smaller angular scales than the dipolar anisotropy. This time the anisotropy has been caused by regions of the Universe which are denser than average. A denser region will decrease the temperature of the emission, because it will gravitationally redshift the radiation more than a less dense volume of the cosmos. Thus, these anisotropies are measures of the distribution of matter in the Universe at the epoch of decoupling.

Figure 10.2. An artist's impression of the Cosmic Background Explorer (COBE), a satellite dedicated to the study of the cosmic microwave background radiation, launched into Earth orbit in 1989. Its measurements confirmed that the spectrum of the background radiation is indeed characteristic of a black-body source at an absolute temperature of 2.7 K, verifying a major prediction of the Big Bang. (Photograph reproduced courtesy of NASA Goddard Space Flight Centre.)

The COBE all-sky map was processed to remove the dipole and the effects of our Galaxy. As a result of the image processing and the faintness of the fluctuations, it was impossible to tell initially whether the patches of light and dark were real features or simply noise in the data. A statistical analysis of the all-sky map revealed that the pattern was not simply random (see Figure 10.3). Despite the image being akin to a static-drenched television picture, some traces of the underlying structure were undoubtedly visible. Besides, a Princeton University balloon experiment to measure the background radiation also discovered ripples similar to the COBE ripples. A correlation between the two data sets revealed that some of the COBE

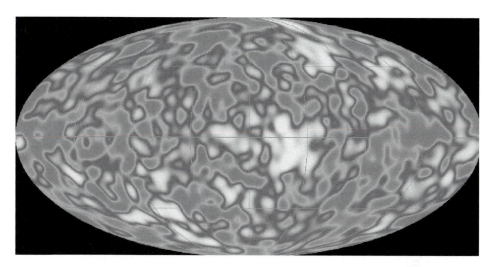

Figure 10.3. COBE ripple anisotropies. If the dipolar anisotropy (Figure 10.1) is modelled by computer and subtracted from the COBE all-sky microwave maps, and if the effects of our Galaxy are taken into consideration, the above remaining map displays fluctuations in the temperature of the cosmic microwave background at the level of 1 part in 100,000. This is the first detection of the 'seeds' of today's large-scale structure. With this image it is impossible to tell if this represents the distribution of matter in the Universe at the point of decoupling. However, sophisticated image analysis has shown that the patterns are not simply the product of random noise. (Photograph reproduced courtesy of COBE Science Working Group, NASA, GSFC, NSSDC.)

features were also detected by the balloon measurements. Recent processing of the COBE data, which used additional observations made by the satellite, to increase the signal-to-noise ratio, actually pinpointed hot and cold spots, which correspond to matter fluctuations at the time of decoupling (see Figure 10.4, colour section).

The fluctuations are very faint, typically possessing $\Delta T \sim 10^{-5}$. This in itself, regardless of distribution, placed the most severe constraints yet on the process of structure formation. The instrumentation on COBE observed the background radiation with an angular resolution of $7°$, so that the fluctuations it observed correspond to extremely large structures, some of which are far larger than any structures so far observed in the Universe today. This raises an interesting question: are luminous galaxies really a good tracer to the matter distribution of the Universe? We will consider this later in the chapter when we develop the idea of dark matter in the cosmos. For now, we return to the fluctuations in the microwave background, and what caused them.

To see the kernel of what became a supercluster requires a microwave telescope capable of resolving the microwave background into half-degree 'blobs'. This is probably the smallest scale on we should expect to see structure in the microwave background. The perturbations which became individual galaxies are probably beyond our observation forever because of the time the Universe took to complete the decoupling of matter and energy.

In our previous discussion we have always referred to this event as if it took place very suddenly. With reference to 15,000 million years of cosmic history, it probably did. However, the Universe did not evolve from being completely ionised to totally neutral overnight. Instead, this must have taken place over a period of time. This means that the surface of last scattering (as the place of origin of microwave background is sometimes known) is a misleading name. Instead of a two-dimensional surface, it must be three-dimensional. Therefore, our line of sight through this transition region of space must include regions of both over-density and under-density, which will tend to cancel each other out. The first anisotropies will therefore occur on angular scales which correspond to the width of the last scattering surface (see Figure 10.5). At present, this is assumed to be a linear distance corresponding to an angular scale of 0.1–0.5°, so primordial cluster fluctuations, and perhaps even groups, should be visible in the microwave imprint.

There is, however, one major caveat that we must always bear in mind when studying the cosmic microwave fluctuations. The theory assumes that the matter in

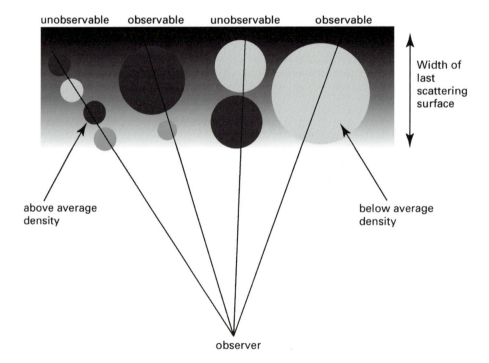

Figure 10.5. Visible fluctuations in the cosmic microwave background radiation. The depth of the last scattering surface affects the scale of the smallest detectable fluctuations. Only those fluctuations equal in size to the depth of the surface will be detectable.

the Universe following this event remains neutral and tucked up in concentrated lumps such as galaxies – but there is evidence that this may not be entirely correct.

10.3 THE ERA OF REIONISATION

In Chapter 8 we discussed the Lyman-α forest and the way chemical elements in space – along our line of sight to distant quasars and galaxies – imprint themselves on spectra. From these spectra it is possible to determine not only the element present in space, but also its ionisation state. From these observations, which stretch between redshifts of 1.7 and 5, it can be seen that a significant proportion of intergalactic gas is ionised. Although the Universe is still optically transparent (because we can see through it) the implication is that there are a lot of errant electrons moving through the cosmos and colliding with photons of the cosmic microwave background. This will blur the primordial fluctuations and, if the electrons are present in sufficiently large numbers, would actually introduce secondary fluctuations which we are currently misinterpreting as ancient in origin. This is the Sunyaev–Zel'dovich process, named after the two astronomers who studied it and presented the idea that it would affect the spectrum of the microwave background, especially in the direction of a rich galaxy cluster. Essentially, it is an inverse-Compton scattering process (described in Chapter 2) in which the interacting electrons possess greater kinetic energies than the microwave background photons which they are scattering (see Figure 10.6).

The observations imply that this ionisation process must have taken place before redshift 5, during the epoch of galaxy formation. One mechanism for the reionisation is the formation of the hypothetical Population III stars, which would have condensed before galaxies had time to accumulate mass. They would have produced large amounts of ultraviolet radiation which readily ionised atoms.

The same ionising process by high-mass stars produces the emission nebulae associated with star-forming regions today. In Orion, for example, the entire nebula is being ionised by the ultraviolet radiation from just one star – θ^1C Orionis – which preserves its surroundings in a constant state of ionisation (see Figure 10.7.). When an electron recombines with a proton to reform a neutral hydrogen atom, it emits its excess energy as one or more photons, which is what causes the nebula to glow and to be seen. However, no sooner has it reformed, than a new ultraviolet photon, emitted by the star, collides with it and reionises it.

The enormous peak in the rate of star formation that has been implied by the submillimetre observations (discussed in the previous chapter) points to the era of reionisation. At that time, the Universe must have been full of titanically large emission nebulae which were gradually ionising the gaseous content of space (see Figure 10.8, colour section).

Whilst some astronomers investigate these effects which alter our perception of the cosmic microwave background's primordial anisotropies, others ponder the primordial fluctuations themselves. In particular, they consider their origin. Again, the COBE data present us with our first clue as to their origin. At the point of

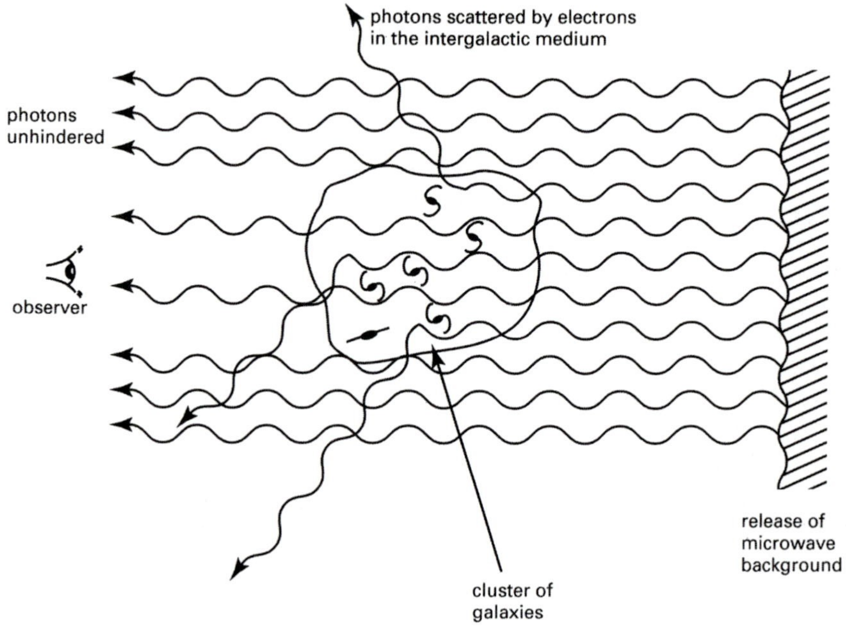

Figure 10.6. The Sunyaev–Zel'dovich Process, involves the scattering of the cosmic microwave background radiation by intergalactic electrons. This is particularly prevalent in clusters of galaxies in which the hot, intracluster gas is in a highly ionised state. Careful observation can reveal 'holes' in the cosmic microwave background, where cluster gas has scattered the background photons.

decoupling, the Universe was approximately 300,000 years old. This means that only areas of space 300,000 light years in diameter would be causally connected, because this is the maximum distance that light could have travelled in that length of time. A diameter of 300,000 light years at the distance of the microwave background's emission corresponds to an angular diameter of 1°.

The COBE fluctuations possess an angular diameter of 7°, which corresponds to a linear diameter some seven times the horizon distance of the Universe at decoupling. The fact that fluctuations exist on scales far greater than the horizon distance at decoupling strongly suggests that they originated before the epoch of inflation at a cosmic age of 10^{-35} seconds. Had this not been the case, then a fluctuation could not have coherently grown so large. One would have expected adjacent regions, outside the horizon distance, to have evolved independently, and therefore not possess the same temperature. When an average of all these individually evolved areas was taken by the 7° beam width of COBE, the fluctuations should have been completely washed out. But this is not the case, indicating that what became the large-scale fluctuations began before inflation. In Chapter 3, the concept of the Heisenberg uncertainty principle (equation (3.3)) was introduced, and in Chapter 4 it was explained how, according to this principle, energy could be borrowed from the

Figure 10.7. The Orion nebula. Virtually all of the emission from this nebula is caused by one star – θ^1C Orionis. This star is part of the quadruple star system known as the Trapezium, but it cannot be seen in this image as it is situated in the bright central region of the nebula. (Photograph reproduced courtesy of Bill Schoening/AURA/ NOAO/NSF.)

vacuum and transformed spontaneously into a pair of particles, providing that it annihilated within its Compton time. At 10^{-35} seconds, the energy density of the Universe was so high that quantum fluctuations were occurring at all energy levels and on all scales. When the Universe inflated due to the separation of the GUT force, these infinitesimal quantum fluctuations were suddenly blown up to encompass macroscopic regions many times the size of the observable Universe at that epoch. Much smaller-scale fluctuations, which then became galaxies, may have been formed in a similar way or at a later epoch before (or even after) decoupling.

10.4 ORIGIN OF THE PRIMORDIAL FLUCTUATIONS

In order to answer the question of what caused the primordial fluctuations to occur, we have to re-examine the theory governing the very early Universe. Chapter 5 described the Universe's history up to the release of the cosmic microwave background radiation. It described how the constituents of the early Universe could be modelled, by treating them as an ideal fluid, subject to random motions and processes. From our consideration of the fluctuation scales detected by COBE, we have deduced that they must have initially formed before the epoch of inflation. There is a deep suspicion among cosmologists that the fluctuations were actually present when the Universe left the Planck era at an age of 10^{-43} seconds. If this is the case, then the fluctuations are almost certainly remnants from the forging of the physical laws and the creation of mass, energy, space and time. We shall return to this point later.

It is assumed that fluctuations occurred on all scales, and so the ideal way to characterise this type of behaviour is to reduce it by Fourier analysis and treat it as a power spectrum. The power spectrum describes the amplitude, A, of the fluctuation which occurs on a scale, k, at a time, t. It takes the form:

$$P(k,t) = Ak^n \qquad (10.2)$$

n is the spectral index, with a value dependent on the assumed type of primordial spectrum. Although nothing in this equation explicitly states that it should be constant over the entire range of k, this is usually assumed. The simplest power spectrum corresponds to a totally random distribution of matter throughout space. This is termed a white-noise spectrum, and is denoted by $n = 0$.

In order to investigate this, let us assume that the matter content of the Universe can be divided into particles of equal mass, and that these particles are then distributed randomly throughout space at 10^{-43} seconds. The number of particles is therefore directly proportional to the mass, M. If the average number of particles in a volume element is N, then the exact number in any specific element may vary by

$$\delta N = \sqrt{N} \qquad (10.3)$$

This comes simply from the Poisson distribution for random events. It follows that, at any particular point in spacetime, the fractional variation can be as much as

$$\frac{\delta N}{N} = \frac{\sqrt{N}}{N} = \frac{1}{\sqrt{N}} \qquad (10.4)$$

which in turns leads to a root-mean-squared variation in the mass, σm:

$$\sigma m \propto N^{-1/2} \qquad (10.5)$$

The basic idea is to then observe how these evolve and grow through gravitational interaction with each other. A white-noise spectrum (which it is the simplest to visualise) unfortunately runs into problems because the fluctuations it causes collapse too early. It actually leads to a 'chaotic' cosmology in which the collapsing regions of space would play havoc with the isotropy of the microwave background and hamper the nucleosynthesis of helium.

Alternatively, if we assume that fluctuations can be present in an otherwise homogeneous distribution of mass simply by rearranging the particles within each specific volume element, we arrive at a spectrum which possesses $n = 2$. This leads to

$$\sigma m \propto N^{-5/6} \tag{10.6}$$

This is often known as the 'particles-in-boxes' spectrum, and postulates that the sizes of the boxes should be in the order of the horizon scale at the time during which the fluctuations were created. Although there are some statistical problems faced by this approach – during the analysis of when the Universe was very young – its main difficulty is that it fails to predict the growth of fluctuations into *bona fide* galaxies in time for us to observe (or even exist in) them!

Finally, we turn our attention to a spectral index, $n = 1$, the Harrison–Zel'dovich spectrum:

$$\sigma m \propto N^{-2/3} \tag{10.7}$$

Having met with failure by trying to interpret the primordial fluctuations as events which occur inside horizon scales, this time the fluctuations occur outside the horizon distance and then cross into an influential range as the original horizon grows with time at the speed of light. Remember that the theory states that the Big Bang took place not in one single place at a single time, but everywhere at the same time. Therefore, at the Planck time many places in the Universe were not causally connected. As the message of their existence propagated outwards in concentric spheres at the speed of light, however, more and more regions were able to communicate with each other through the exchange of particles. Although the Universe was expanding, it was not expanding faster than the speed of light, and so the causally connected Universe – which became our own – continued to become bigger and bigger. As it grew and encompassed more regions of spacetime, so external fluctuations crossed the horizon and began to effect it. One of the most interesting properties of the Harrison–Zel'dovich spectrum is that it predicts scale-invariant fluctuations: the amplitudes of the fluctuations are not correlated with the scales upon which they occur. This is the type of spectrum most cosmologists assume, because it can produce figures which predict the right level of fluctuations in the cosmic microwave background, and can produce galaxy formation on acceptable time-scales. It also happens to be predicted by inflationary cosmological models.

Cosmologists still do not know the fundamental reason for the initial presence of the fluctuations. Most assume that it is a remnant of the quantum gravitational forces which were in flux during the Planck era. Subscription to the Harrison–Zel'dovich spectrum provides an escape from providing an explanation, because different regions of the Universe – beyond each other's horizon distances – would not be expected to display similar characteristics. Having established a preferred power spectrum of fluctuations, regardless of their initial cause, the next step is to determine the way in which they grow.

10.5   GROWTH OF FLUCTUATIONS DURING THE EARLY UNIVERSE

During the radiation-dominated era, gravity was not the dominant force in the Universe. Instead, it was superseded by the pressure caused by photon collisions. Therefore, any fluctuations which did occur could not grow through the accumulation of more matter, because the radiation pressure smoothed them away.

As the Universe expanded, the smoothing action of the photons became less and less efficient. In Chapter 4 we defined the beginning of the matter-dominated Universe as being when the energy density of space – and hence its expansion – was governed by its content of matter, rather than by its radiation. At this point, gravity began to take control, and shaped the structures in the Universe. Even though matter domination began about 100 seconds after the Big Bang, radiation pressure still restrained the growth of the structures quite effectively until the Universe reached an age of 10,000 years. Interactions between matter and radiation continued, however, inhibiting the growth of fluctuations so that the structures grew only slowly, until 300,000 years. At this point, interactions between the radiation and matter ceased, and the cosmic microwave background was released.

The precise way in which a fluctuation grows before the decoupling of matter and energy depends upon whether it includes an increase in the density of radiation, or whether it is just the matter content which is affected. Thus, two types of density perturbation are possible in a Universe containing matter and radiation. They affect whether the entropy of the early Universe varies from place to place or remains constant throughout space. Entropy is the measure of a system's ability to undergo spontaneous change. It can also be thought of as a measure of the system's ability to do outside work. It is thermodynamically defined as:

$$dS = \frac{dQ}{T} \tag{10.8}$$

where dS is the change in a system's entropy based upon the absorption of a minute quantity of heat energy, dQ. The system itself, throughout this reaction, remains at the temperature, T.

An adiabatic fluctuation is one in which both the density of matter, ρ_m, and the density of radiation, ρ_r, are perturbed from the average density, $\Delta\rho$, in such a way that the entropy of the perturbed region remains constant with its surroundings. This condition is met when matter and radiation perturbation occurs in the following way:

$$\left(\frac{\Delta\rho}{\rho}\right)_r = \frac{4}{3}\left(\frac{\Delta\rho}{\rho}\right)_m \tag{10.9}$$

The second type of perturbation – isothermal fluctuation – occurs when the matter is the only component of the Universe to be perturbed. The fundamental differences between these two types of density enhancement are their effects on the spacetime continuum. An adiabatic fluctuation alters the curvature of the spacetime continuum, whereas an isothermal one does not.

The type of fluctuation is only a concern during the radiation-dominated Universe and the period of time up to the decoupling of matter and energy. If the fluctuation is not large enough (does not contain the Jeans mass defined in Chapter 9), it is the action of the radiation which dissipates the fluctuation. Following the decoupling, the radiation content of the Universe (observed today as the cosmic microwave background radiation) and matter evolved separately and no longer influenced each other to any great degree. At the point of decoupling, the type of the initial fluctuations determines on what scale galactic structure will first emerge.

In the case of adiabatic fluctuations, the top-down scenario is proposed, in which the first structures to form are clusters and superclusters. The reason for this is that the fluctuations attempt to confine an over-density of radiation, which then diffuses. This diffusion process is characterised by the collision of photons with the 'encasing' matter which, in tandem with the external buffeting received by the fluctuation, increases the kinetic energy of the particles and smooths out the fluctuation. Only very large fluctuations are capable of surviving this process, because the time take for the radiation to diffuse out of them is much longer than the time to decoupling. At the point of decoupling, the only fluctuations capable of surviving are those which contained 10^{15} or more solar masses. Thus, structures on the size of clusters and above are the first to emerge. Computer simulations show that these then collapse into pancake-like structures and fragment into individual galaxies. This would naturally lead to large-scale structures in which sheets and filaments of galaxies surround large voids. Whilst there is evidence from galaxy maps that the real Universe is distributed in this way, the model presents problems. It predicts too much large-scale structure and excessive fluctuations in the cosmic microwave background radiation.

Isothermal fluctuations lead to a bottom-up approach, in which smaller associations such as galaxies, or even globular clusters, form first. The larger constructions – giant galaxies, clusters and superclusters – later form from gravitational instabilities caused by the smaller aggregates of matter. This hierarchical clustering would lead to self-similar clustering patterns. Essentially, the superclusters of galaxies should resemble the clusters of galaxies which, in turn, should resemble the galaxies. Although the concept of hierarchical clustering would appear to be broadly correct, observations of the surrounding Universe indicate that there are differences in the properties of galaxies, clusters and superclusters. These differences indicate that different physical processes are at work on these different scales. For example:, on the scale of galaxies, gas dynamics would seem to play an important role in the shaping of spiral galaxies: on the scale of superclusters and above, filamentary structure – such as that predicted by top-down scenarios to be abundant – is also present.

Both models also suffer from the common problem mentioned earlier: the contrast in density between a galaxy and its surroundings is colossally different from the level of the fluctuation required to begin its gravitational collapse. It is so different, in fact, that more matter than appears to be present needs to be added in order to cause the structures in the Universe to form in time for us to observe them. Our estimates of the matter content of the cosmos fall short of that required by these models to build galaxies and clusters. If the deficit is made up by assuming that there are large quantities of normal baryonic matter (neutrons and protons) which we simply cannot see, then the amounts of helium synthesized in the early Universe (see Chapter 5) do not accord with observations. Unfortunately, the inescapable conclusion is that the Universe is more complicated than a simple interplay between baryonic matter and radiation.

We need somehow to introduce particles which would provide extra gravity to facilitate the building of structures in realistic time-frames. At the same time, we need to introduce some kind of smoothing effect to produce the observed sheets and voids. Whatever these particles are, all other interactions – which would naturally affect the baryonic component of the Universe – cannot occur, because they are observed not to occur. It is quite difficult to account for all of this, but modern particle physics has done its best!

10.6   DARK MATTER

During the last half-century there has been growing evidence that a substantial fraction of the matter content of the Universe is in forms which cannot be directly observed. World-renowned astronomer Jan Oort produced the first evidence that forms of matter were hidden from our view, when he conducted a study of stellar motions in the vicinity of our Sun. When considering that, in our solar neighbourhood, we are embedded within the outer disc region of the Galaxy, Oort discovered that most stars have a velocity component out of the plane of the disc. This perpendicular motion would be stopped and eventually reversed by the mass of material in the Galactic disc – exactly in the same way that a ball thrown upwards from the Earth is slowed by gravity and is eventually pulled back down. This would cause an oscillation, as the stars plunged back through the centre of the disc and out of the other side. Thus, as they orbit the centre of the Galaxy, stars also oscillate up and down out of the disc. However, a single cycle of these oscillations lasts for thousands of years, and they are therefore unobservable within the span of a human lifetime. A study of stellar distances from the Galactic plane and the velocities possessed by those stars does allow an estimate for the gravitating mass in the Galactic disc. Oort discovered that there had to be more mass in the solar neighbourhood than could be accounted for by stars alone. The mystery was partially resolved with the advent of radio astronomy in the 1950s. These new instruments proved that interstellar space is home to vast quantities of dust and gas – which at first had not been suspected. Nevertheless, doubts remain about whether this constitutes the total hidden mass.

The next pieces of evidence came from studies of the rotational rates of stars in other spiral galaxies. By assuming that the luminous portions of a galaxy trace the mass distribution, a spiral galaxy should be a classic case of a Keplerian orbital system (similar to the Solar System) which has a central concentration of mass (the Sun or the galactic nucleus), around which other much smaller bodies orbit. In the Solar System this results in the furthest objects from the central body having the smallest orbital velocity. Rotation curves of galaxies, however, show that this is not the case. Orbital velocities for spiral galaxies reach a plateau value, and remain there for the width of the disc. This behaviour would indicate that the spiral galaxy is surrounded by a large spherical cloud of matter. In Chapter 9 it was stated that evidence for large haloes of gas has been found around external galaxies, but in many estimates the mass of gas is smaller than the mass inferred from the rotation curves, and hence the presence of another dark component of matter is suspected.

Extending our observations to the scale of clusters of galaxies, we find that the individual velocities of galaxies within these clusters should imply that they cannot remain together, if the only mass present is that which can be seen. Because clusters of galaxies can be seen throughout the Universe, it can be inferred that they are stable structures; and to be stable they must contain much more mass than is obvious. Haloes and the suspected propensity of tiny, faint dwarf galaxies may help to make up this shortfall, but other more exotic matter components are still necessary.

10.7   HOT DARK MATTER AND MASSIVE NEUTRINOS

So, what is this exotic form of matter which we cannot see but can detect through gravity? There are two possible answers to this question. For our first we will use dark matter candidates which are known to exist: neutrinos. Conventionally it has always been assumed that neutrinos possess no mass and travel at the speed of light. They were originally postulated to exist in 1930, when energy deficits were discovered in certain sub-atomic reactions. Knowing that energy could neither be created nor destroyed, Pauli – the man responsible for espousing quantum theory's exclusion principle – proposed that an unknown particle was taking part in the reactions and carrying off the excess energy. Theoretical considerations were made and its properties deduced. Neutrinos take part only in reactions which involve the weak nuclear force. They were shown to be extremely reluctant to take part in reactions. Neutrinos need to be incredibly close to atomic nuclei before a reaction can take place, because the weak force operates over such a short distance. For a neutrino, the separation between atoms is so large that they can literally find an unhindered straight path through matter in the same way that a human being can walk through a sparsely wooded forest. During the time taken you to read this sentence, millions of neutrinos have passed through you, this book and the entire planet Earth! Neutrino theories have been largely corroborated with the invention of the neutrino detector – an enormous swimming pool-like device, filled with water or some other liquid such as tetrachloroethene, which catches one or two neutrinos out of the countless billions which pass through it! As for the neutrino's mass: for many years, particle physics could determine only that it is very, very small, and therefore probably zero. It was always possible that this was not the case, and if a neutrino has even a minuscule mass, then they could be the major constituent of the dark matter in the Universe, because they populate the Universe in such vast quantities.

A recent development, however has considerably increased the chances of determining a definite mass for the neutrino. A team of 120 Japanese and American physicists are collaborating on a project known as Super-Kamiokande – an enormous 50,000-tonne tank of water, surrounded by light-sensitive detectors which record the faint flashes of light that are produced when a neutrino makes an exceedingly rare interaction with one of the water molecules. Super-Kamiokande is installed 2,000 feet beneath a Japanese mountain to prevent as many other stray particles as possible from entering it. By analysing the results of the detections, the team of physicists realised that there was an obvious deficit in the number of neutrinos detected. The only way to solve the problem was to believe that the neutrino had a low mass. In Chapter 3 it was stated that the Standard Model of particle physics predicts three different varieties: the electron neutrino, the muon neutrino and the tau neutrino.

Physicists are now beginning to realise that not only does the neutrino have a low mass, but that it can change from one type of neutrino into another, in a more or less spontaneous fashion. The rate at which it changes its mass depends upon its mass. The Super-Kamiokande experiment has shown that the mass of a muon neutrino is likely to be about 0.1 eV. A normal electron has a mass of 500,000 eV. According to theory, neutrinos outnumber electrons and protons by about 600 million to 1 – 300 of them in every cubic centimetre of the Universe! In this case, the amount of mass in the Universe virtually doubles, simply because the neutrino possesses a tiny mass.

This discovery was swiftly followed by the announcement that the University of Minnesota will extend the capability of their world-renowned particle physics research centre, Fermilab, to investigate these neutrino phenomena. A new underground laboratory is to be excavated, and it will be ready for work by 2002. The new experiment will generate a pulse of muon neutrinos and fire it towards a target. A new pulse will be generated every 1.9 seconds, and each pulse will contain around 300 trillion neutrinos. The neutrinos will be detected by an amazing detector consisting of a series of steel plates, comprising 10 million pounds of steel. The chances of any single neutrino interacting with it are remote; but even so, the international group of scientists responsible for this instrument expect to detect only one neutrino for every 1,000 pulses. To compensate for this minuscule detection rate, they expect to fire pulses continuously for nine months of every year; and they believe that after four years of operation they will have enough detections to calculate the chances of a neutrino spontaneously changing from a muon flavour into another flavour. This will then allow them to calculate the mass of the muon neutrino, and hopefully confirm the Japanese result.

Neutrinos are relativistic particles; that is, they rush around the Universe at velocities very close to the speed of light. They are therefore classified as hot dark matter. They ceased to take part in the thermal evolution of the rest of the Universe a long time before matter and radiation decoupled. In fact, neutrino decoupling is expected to have taken place at about one second after the Big Bang, and since that time they have hardly reacted with matter at all. Their very high speed causes them to resist clumping, and their inherent gravity has a smoothing effect on the distribution of matter, pulling apart the density fluctuations which occur in the baryonic matter.

The hot dark matter scenario of galaxy formation is therefore similar in outline and consequence to the older adiabatic scenario which led to top-down structure formation. Whilst the determining property of the matter–radiation scenarios was the type of the initial fluctuation, this is not the case in dark matter models. Whilst two principal fluctuation modes are still possible, the overwhelming factor is the type of dark matter present in the Universe. It is this which determines the subsequent evolution. In the first type of fluctuation, all three of the Universe's components are perturbed: radiation, matter and dark matter. This is an adiabatic fluctuation which causes a perturbation in the spacetime continuum. In the second type of fluctuation – isocurvature fluctuation – the net-energy density remains constant despite the density fluctuation, and so the spacetime continuum is unaffected. In most of today's cosmological models, the more general case of the adiabatic fluctuation is the only one to be considered.

Hot dark matter leads to top-down structure formation in which cosmic pancakes condense to form clusters and galaxies (see Figure 10.9). In common with their matter–radiation counterpart, these models again run into problems with their predictions of too much large-scale structure. In the real Universe, sheets and voids and filaments are there, but are simply less numerous and smaller than those of the simulations.

The 'washing out' of small-scale fluctuations occurs because of the neutrinos' great speed, so the other possible candidate must be a type of particle which possesses mass but is not relativistic. To distinguish them they are referred to as cold dark matter. Unfortunately, unlike hot dark matter, there is no known candidate particle. Numerous theoretical possibilities exist, however, and are tied into the evolution of the Universe at very early times. The cold dark matter candidates are best thought of as cosmic relics. During the earliest stages of the Universe they served their purpose, influenced reactions and helped to space our cosmos. Now, however, they simply litter the cosmos and do very little except influence baryonic matter with their gravity.

10.8   COLD DARK MATTER

Like the neutrinos, cold dark matter decoupled from the rest of the Universe long before matter and radiation departed thermal equilibrium. It began clumping because the cold dark matter was slow-moving and unreactive with the rest of the Universe. When the cosmic microwave background was released and radiation ceased to support the baryonic matter, it could fall quickly and efficiently into the ready-made sites of gravitational potential, where the dark matter had clumped together. Thus, cold dark matter models follow the bottom-up scenario of structure formation (see Figure 10.10).

Just what constitutes the cold dark matter is a problem which has been vexing astronomers and particle physicists for more than a decade. In Chapter 4 the description of the Big Bang described how the forces of nature are thought to have been unified at earlier times, when the particles and radiation were at greater

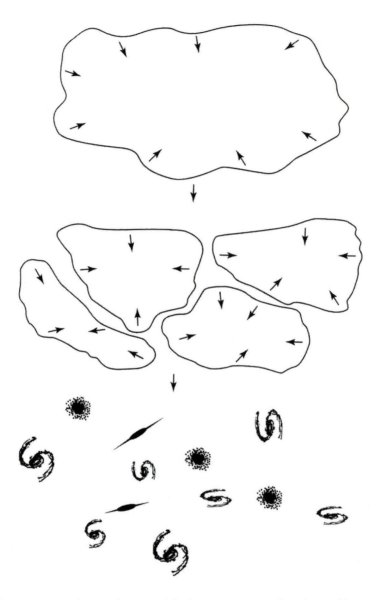

Figure 10.9. In the top-down model of structure formation, large objects collapse to form cosmic pancakes. These then fragment, and galaxies form.

energies. The unification of the weak nuclear force and the electromagnetic force has been proved by high-energy experiments in particle accelerators. The theorised unification of this electroweak force and the strong nuclear force has yet to be shown experimentally but many still believe in its validity. Known as grand unification, it is during this era of cosmic history that baryons are expected to form. As a by-product of this process, one cold dark matter candidate is produced: the axion.

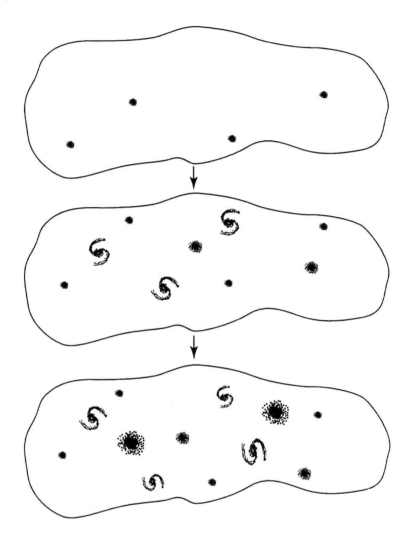

Figure 10.10. In the bottom-up model of structure formation, small objects form first, followed by larger and larger objects. Eventually the Universe is filled with large objects, such as superclusters, which are conglomerations of the smaller objects.

During the advent of quantum and particle physics in the first half of the twentieth century, the available evidence seemed to suggest that all the particles of nature followed three simple symmetries (or invariances, as they are usually called).

C invariance. Particles and antiparticles obey the same physical laws. C denotes charge.

P invariance. The laws of physics are exactly the same if the quantum spin states of the reacting particles are reversed. This is like replacing the reacting particles with mirror images of themselves. P denotes parity.

> T invariance. The laws of physics are unchanged if the direction of time is reversed. T denotes time.

In 1956 the violation of these invariances was predicted by physicists Tsung-Dao Lee and Chen Ning Yang. Their work indicated that the weak nuclear force violated the P invariance. This has been corroborated for neutrinos, the signature particle of the weak force. Neutrinos always display an antiparallel sense of spin in relation to their motion, which is a clear violation of parity because one would expect there to be equal numbers of parallel and anti-parallel spinning neutrinos. And even worse than this, neutrinos also violate the C invariance! This is because anti-neutrinos always display parallel spin vectors in relation to their momentum vectors. Considered overall, however, this means that the weak nuclear force does display a combined CP invariance: if the particles of matter in the Universe were replaced by antiparticles and their spin states were reversed, the Universe would still evolve in the same way that ours has evolved.

In the mid-1960s, evidence emerged that the decay of particles called kaons did not follow CP invariance. Finally, as a last resort, particle physicists still believe in CPT invariance, especially since mathematical constraints have bolstered its possibility. For a Universe to evolve in the same way as ours, it would need to have all particles of matter replaced with antimatter, and their spin states reversed; and in addition to this, time would have to flow backwards!

Individually, therefore, the invariances can be broken. In the high-energy environment of the early Universe, so many reactions took place that even statistically improbable events occurred in abundance. It would be the perfect place to study these invariance violations. In the process of breaking the invariances – particularly the C invariance – the excess of matter over antimatter can be built up. T invariance can also skew the ratio of baryon creation to baryon destruction in favour of a matter Universe. Some versions of this baryosynthesis model involve the creation of axions. An axion is expected to have a small mass but to be present in such large numbers that it could contribute a considerable amount of mass to the Universe.

Another leading idea about the nature of the cold dark matter points to its creation even earlier than baryosynthesis. In trying to unify gravity with the grand unified force, theorists really need a quantum theory of gravity. The first steps have been taken, however, with supersymmetric theories. In these mathematical constructions, the dividing line between fermions and bosons is removed, and they are able to change into one another. As a consequence, every boson must have a supersymmetric fermion counterpart. The names of these hypothetical particles are constructed by taking the original particle's name and adding a prefix or a suffix. The prefix 's' is reserved for the fermion counterpart of a boson, – a selectron, a squark. The suffix of a fermion is changed slightly to 'ino' to define its boson counterpart. Thus, photinos and gravitinos abound. It is these final two super-symmetric counterparts in which we are primarily interested, because, according to the theory, they are the only two which are stable. We shall concentrate on the photino, because at least it is known that photons exist; and gravitons still reside in the 'unproven' file. The photino is theorised to have a mass of 10–100 times that of a proton. It is therefore a very heavy particle, and as a consequence is slow-moving and weakly interacting. This has led to its being generically termed (with others like it) WIMP – weakly interacting massive particle!

There is, as yet, no evidence of the existence of these WIMPs, but particle accelerators are gradually being upgraded to bring their energy thresholds within grasp. At the time of writing, the European laboratory for particle physics, CERN, is embarking on a search for these hypothetical particles using its powerful Large Electron Positron Collider (LEP). Physicists are hopeful that by recreating the conditions which are predicted to have prevailed a mere 10^{-10} seconds after the Big Bang, they may be able to glimpse the supersymmetric particles (known as 'sparticles' to their pundits) being created and decaying.

Theoretical work on cold dark matter and hot dark matter is now awaiting experimental verification and tighter observational constraints. The proof of a neutrino's mass and the discovery of a WIMP are both of top priority. Without them, theories can be built on theories – but this is really like building paper houses. Unless they are firmly rooted in observational or experimental evidence, they run the risk of being blown away.

We defined the edge of our Universe as being the observation boundary created by the release of the cosmic microwave background – just as the Sun's boundary is defined by the release of light at the photosphere. With neutrino detectors, however, we can peer into the very heart of the Sun and observe the fusion reactions taking place at its core. If we possessed sufficiently large and sensitive neutrino detectors, then the cosmic neutrino background would also be revealed.

In Chapter 5 we discussed neutrino decoupling, and stated that it occurred about 10^{-5} seconds after the Big Bang. If a neutrino detector of sufficient power were to be constructed, this would provide cosmologists with an opportunity to observe the conditions of the Universe just 0.00001 seconds after creation itself! All of our theories of baryosynthesis, supersymmetry and primordial density fluctuation could be severely tested, and our understanding would take a massive leap forward.

As physicists struggle with quantum theories of gravity, they too must have an ultimate goal in mind. By successfully proving the existence of the graviton – presumably with the a sufficiently large and sensitive graviton detector – the Universe could be studied as it existed at the end of the Planck era, 10^{-43} seconds after the Big Bang. This was the point at which the gravitons are theorised to have decoupled. Observational cosmology is really only just beginning!

11

The fate of the Universe

Throughout this book there have been frequent references to the age and size of the observable Universe. According to current estimates, the age could be anything between 10 and 20 thousand million years, and we have considered 15 thousand million years to be an acceptable estimate. The age of the observable Universe is directly linked to its size; about 10–20 thousand million light years. A quantity that we have yet to fully consider, however, is the mass of the Universe. Throughout most of this book it has been tacitly assumed that the matter in the Universe is the luminous material in stars and galaxies. The work described in the previous chapter, however, has shown that the actual content of the Universe is rather more complicated to define. If there are vast quantities of dark matter (whether hot or cold makes no difference to this argument) then astronomers are faced with the conclusion that baryonic matter may not be the best blueprint with which to study the Universe. It would be the equivalent of an ecologist studying Earth's biosphere, but considering only human beings and their impact. If the ecologist was to ignore all of the other living species on the Earth he could not hope to develop the correct model of our planet's ecological system. Astronomers today are faced with the prospect that they have been studying the Universe using the most restricted of data sets and missing the 'big picture'. To use another analogy: they have been peering at a city during the night and seeing only the windows in which lights are burning, rather than the skyscrapers to which those windows belong.

11.1 THE DENSITY OF THE UNIVERSE

Psychologists reading this book may have noticed an interesting trend among astronomers and cosmologists in that, whilst they pretend to understand the grandest concepts in the Universe, they are forever inventing new particles, ideas and units to make it comprehensible. When distances in miles became far too large to conceptualise, the light year was formulated. When the plethora of sub-atomic particles became too dizzying, they developed the Standard Model, containing quarks, to reduce the number of particles. The same will occur in our discussion of the mass of the Universe: we will reduce everything to a single critical number.

Instead of discussing the mass of the Universe, we shall use a related quantity – density – because this takes into account the expansion of the Universe. Chapter 5 introduced the Friedmann equation (5.29), which related the growth of the scale factor to the interplay between the Universe's potential gravitational energy and its kinetic energy. These quantities were related to the density, in the case, of the potential energy, and to the rate of the expansion, as quantified by the Hubble constant, in the case of the kinetic energy. Three possible results were briefly introduced: the open, 'flat' and closed scenarios. Which one of these Universes we live in depends upon the values of the Hubble constant and the density. The energy constant, k, in the Friedmann equation is the diagnostic feature which allows us to distinguish between these alternatives. Positive values of k correspond to open Universes. In the case of an open Universe the amount of matter it contains is too small to create enough gravity to halt its expansion, and it will expand forever. The relative rate of the expansion is determined by the magnitude of k.

11.2   OPEN UNIVERSES

As the open Universe expands, the clusters of galaxies will become ever more distant from one another. The continued expansion of the Universe will render the superclusters more extended and tenuous, but the individual clusters should remain relatively undisturbed, because their force of self-gravity is strong enough to overcome the Hubble expansion. As stated in Chapter 7, the haloes of matter which have been observed around some galaxies may still be feeding these vast star-forming systems with raw materials. Such galaxies – predominantly spirals and certain types of irregulars – will continue to form stars from their gas reserves for many billions of years. Those stars may then continue to shine for many more billions of years, depending upon their size. The amount of raw material in the Universe, whilst vast in quantity, is not infinite. Stellar activity must cease at some point when the Universe is depleted of useful gases. This is the stage which is expected to be reached after it exceeds an age of about 10^{12}–10^{14} years – approximately 66–6,666 times the current age of the Universe.

Although the Universe will no longer contain bright, hydrogen-burning stars, galaxies in the broadest sense should still be identifiable as large conglomerations of mass. They will contain stellar remnants – such as white dwarfs, neutron stars and black holes – orbiting around the nucleus, in much the same way as did the luminous stars. As the Universe ages still further, the number of close encounters between these stellar remnants will become significant. It is important to remember that the frequency of these encounters is not likely to increase, from their present-day scarcity, in this future time, but they receive our attention because nothing more interesting is happening! As a consequence of their relative scarcity, the time-scales on which they influence significant change becomes very large indeed. In the encounters, kinetic energy can be transferred from one remnant to another. Thus, whilst one is given more energy so that it may escape from the galaxy altogether, the other loses energy and falls towards the central regions. The stellar remnants which escape seem destined to wander the depths of intergalactic space for the rest of eternity or until they are captured by some other heavily gravitating body. Those which fall towards the centre of the galaxy will increase the density there, and allow massive black holes to form. If a black hole already exists, it will contribute more mass to it and, as more stellar remnants fall into these central regions, there may be a brief resumption of activity similar to that found in active galaxies.

By the age of 10^{18} years, the Universe will be populated by a very diffuse sea of intergalactic vagabonds and the parent galaxies from which they were ejected. The galaxies themselves will contain a paltry population of stellar remnants dominated by a central, supermassive black hole. Eventually, after about 10^{20} years, the galaxies will be nothing but supermassive black holes. After this, the various clusters of these galactic remains (which were once clusters of galaxies) will gradually coalesce and form even larger black holes, in the same way that the stellar remnants coalesced together. This process will produce cluster remnants which consist of nothing but a single, gigantic cluster-mass black hole. The Universe is expected to be in this state by the time it reaches an age of approximately 10^{30} years.

At this point in the future, two possibilities are allowed for within the scope of modern physics. Which one will actually happen will depend upon whether or not protons will decay into lighter elements. This decay is predicted by certain grand unified theories, which seek to link the strong nuclear force to the electroweak force. It does this by predicting that hadrons (such as protons) can be turned into leptons (such as electrons). The precise mechanism by which this hypothetical process is supposed to take place is as follows. As discussed in Chapter 2, the proton is made of three quarks. If the quarks are thought to be the size of golf balls, the proton they create acts as if it were the size of the Solar System. The three quarks are moving around inside that volume, and attract each other via the elastic-like interquark force. The volume of the proton is defined because, if a quark begins to stray too far, it is pulled back by the other two! Should two of the three quarks contained within the proton move close enough to one another, they will exchange a virtual particle, and the proton will decay into a neutral pi-meson and a positron. The pi-meson will then annihilate itself because it is be composed of an up-quark and its antimatter counterpart.

Assuming for now that this can occur, the lifetime of a proton is calculated to be somewhere in the region of 10^{30}–10^{35} years. When they decay they eventually turn into lighter particles – leptons, which are particles such as electrons, photons and neutrinos. Thus, if protons do decay, atoms will no longer be able to exist, and so any remaining stellar remnants will be dissociated. The Universe will become a dilute sea of leptons, pock-marked by the occasional supermassive black hole. Not even this unpleasant possibility marks the eventual end, however. If Stephen Hawking's ideas about black holes are correct, they can evaporate when quantum fluctuations in the vacuum take place close to their event horizons. When this occurs, a pair of virtual particles is created, one of which falls into the black hole, depleting it, whilst its counterpart escapes and produces the impression that the black hole has evaporated. This incredibly slow process gradually converts all the black holes into leptons after 10^{100} years.

If, however, protons do not decay, then space will still be filled with a few dense stellar remnants such as neutron stars, even after all the black holes have evaporated. It is conceivable that the Universe will eventually reach a position where it is in thermal equilibrium; that is, all its constituents will be at the same low temperature. In this situation, all chemical reactions cease, and the Universe is said to have died the 'heat death'.

11.3   CLOSED UNIVERSES

Another possibility is that the sign of the energy constant, k in the Friedmann equation, is negative. This leads to a Universe which is closed. In cases such as these, the Universe contains so much matter that its gravity can halt the expansion and then reverse it. This will lead to a collapse of the Universe, and a big crunch!

In this scenario, the clusters of galaxies will eventually slow each other down by the force of their mutual gravity, and will then begin to attract each other. Whilst it may appear, at first sight, as if the Universe is running in reverse, it is important to remember that time will still flow forwards; for instance, hydrogen will still fuse into helium in the centre of stars. So, stars will still be born out of collapsing gas clouds, they will live out their natural lives, and will die by becoming either planetary nebulae or supernovae. The only difference will be that the galaxies become closer together rather than farther apart, because instead of an expansion there will be a contraction of the spacetime continuum. The most noticeable change that this will induce is that the distant galaxies will display blueshift rather than redshift, because the electromagnetic wavelengths will become 'squashed' by the collapsing Universe rather than stretched by an expanding Universe.

The other obvious difference will be that the temperature of the cosmic microwave background radiation will appear to increase, because its photons will be blueshifted to higher energies. For the majority of the time, however, conditions in a collapsing Universe will be very similar to that in an expanding Universe. Only during the final billion years of its existence will the events become interesting!

As the volume of the Universe shrinks, the first event will be the merging of superclusters and then clusters of galaxies, which will increase the number of galaxy interactions. The incredibly frail spiral galaxies will be the immediate casualties of this consequence. They will be converted through mergers and interactions into irregular and elliptical galaxies. Giant elliptical systems – similar to the cD galaxies found at the centres of clusters – will form in abundance. Interactions and mergers will become such commonplace occurrences that, probably 100 million years before the end, the Universe itself will simply be a huge collection of stars. There will effectively be no individual galaxies, as they will all have lost their identities in this titanic merger.

As time passes, the individual stars and planets will continue to be pulled together by the force of their mutual gravity. Some of these objects will collide, and as they accumulate more mass and become more dense they will become black holes. These will then grow at a prodigious rate, because the density of the Universe will increase the probability of collisions. If the remaining stars and planets escape the black holes, they will simply evaporate into space. This bizarre fate is caused by the increasing blueshift of the cosmic microwave background radiation. Its temperature will rise so much that it will appear that space itself has reached a temperature of several tens of thousands of Kelvin. The various types of star will all dissolve as space becomes hotter than their surfaces. Eventually, the Universe will reach a stage where atoms will be dissociated, and then baryons will be broken into quarks. In the final few seconds before the final cataclysmic end, the conditions may be very similar to those which existed in the first few instants following the Big Bang. This notion has led some to suggest that the Universe may 'rebound', and that out of the big crunch another Big Bang will occur. Others believe that the Universe will simply return from whence it came.

11.4 'FLAT' UNIVERSES

The dividing line between the open and closed Universe possibilities is a special case known as a 'flat' Universe. This corresponds to the energy constant, $k = 0$. Because it contains just enough matter to stop expanding after an infinite length of time, the fate of a 'flat' Universe would actually be indistinguishable from that of an open Universe. At this point, it is also worth mentioning that a fourth possibility is that our Universe is a closed Universe which also happens to be very long-lived. In this case it will evolve along the lines of an open Universe, but will eventually collapse into a big crunch after the stars have ceased to shine.

With the theoretical cosmologists having defined the various possibilities, it is now the task of observational astronomers to search for the evidence to determine in precisely which type of Universe we live. But this is not easy (as those who have read this book from the beginning will no doubt have already guessed). As stated earlier in this chapter, the two quantities required, in order to define the type of Universe we live in, are the Hubble constant and the density. Chapter 5 discussed the difficulties of pinning down the value of the Hubble constant, and for the sake of argument we have, throughout this book, often used the average value of 75 km/s/Mpc. The final requirement is an estimate of the value of the density; and the range of possible densities can be restricted with some simple deductive reasoning.

The value of the Hubble constant is of crucial importance to these estimates, because it helps to set a lower limit to the amount of mass contained within the Universe. Had this mass been too low, the Hubble constant would have been much higher, and the cosmos would have expanded so much that its constituents would not have been able to group together through gravity. In a Universe such as this, galactic structures would have been unable to form. Conversely, because the Universe is about 10–20 thousand million years old (and still displaying a redshift) the amount of matter it contains also cannot be too great. If the density were very large, the Universe would be expanding only very slowly, or would have already begun its collapse; or, in extreme cases, have already collapsed. Thus, by using little more than some very general observations, we can conclude that the density of the Universe must lie at a point somewhere within a narrow range of values.

In order to quantify this range we must introduce a new concept. If a value for the Hubble constant is assumed, it can be shown very easily from the Friedmann equation (5.29) that the density required to create a 'flat' Universe – as the critical density, ρ_{crit} – is expressed as:

$$\rho_{crit} = \frac{3H_0{}^2}{8\pi G}$$

(11.1)

The actual density of the Universe, ρ, is usually expressed as a ratio of this, Ω:

$$\Omega = \frac{\rho}{\rho_{crit}}$$

(11.2)

Open Universes are represented by $\Omega < 1$ and closed Universes are represented by $\Omega > 1$. From the simple fact that we exist in an expanding Universe which contains gravitationally-built constructions, we can estimate that our Universe lies somewhere in the region $0.1 < \Omega < 10$. Although we have discarded many possibilities, we have not been able to choose which type of Universe we live in. All we can say is that its values are very close to those of a 'flat' Universe. This statement, however, is by itself quite powerful in modern cosmology, because it begs the question: why?

Given that the Universe could have been characterised by an infinite number of different positive Ωs or an equally infinite number of different negative Ωs, why is it so close to the one and only special case? Perhaps the reason that it is so close to the special case is because it *is* the special case: a 'flat' Universe in which $\Omega = 1$. This conundrum has been elevated to the status of a named cosmological problem – the flatness problem, discussed in Chapter 5. It was also explained how the theory of inflation produced a flat, observable Universe, regardless of its initial curvature.

11.5　OBSERVATIONAL TESTS OF INFLATION THEORY

Accepting, for now, that we do indeed live in an inflationary Universe, then we should be able to observationally verify this in a number of different ways. Firstly, because the Universe is expanding we expect the density of galaxies to increase as we look further and further into space. This is a consequence of look-back time, because we see these distant parts of the cosmos as they appeared thousands of millions of years ago when the Universe was smaller and the galaxies were closer together. This technique is known as source counting and, in principle, should be able to determine any curvature possessed by the spacetime continuum.

The curvature of spacetime will distort the source counts away from those expected from the straightforward expansion of the Universe, because the gravitational curvature takes place in a fourth dimension which human beings cannot perceive. Although it is impossible for us to see the curve directly, we become aware of it because it causes a distortion in our familiar three dimensions. In order to appreciate how the curvature of space affects source counts, we can visualise a two-dimensional model Universe. Imagine a sheet of rubber. It is a two-dimensional object, and we chose to deform it in some way through a third dimension. The curvature of the Universe is theorised to be a constant property throughout its volume. Therefore, it can display either positive curvature – in which case our rubber sheet would become a hemisphere – or negative curvature – in which case it would become a hyperbolic saddle-shape (Figure 5.8 illustrates these geometries). On this sheet of rubber there are dots, painted at random, but maintaining a constant density across the rubber sheet.

A two-dimensional being on the rubber sheet cannot perceive that its Universe has been distorted, and still thinks of it and sees it as being two-dimensional. To simulate what this two-dimensional astronomer would see, we have to decide what we need to do to the rubber sheet in order to return it to its two-dimensional form.

In the case of the negatively curved Universe, we could 'pull' the edges up, thus stretching it flat. This stretching would be more pronounced at the edges than at the centre, and so the density of dots would appear to decrease the further away the astronomer looked. In the case of a positively curved Universe, however, the edges would have to be squashed in order to flatten it. This would cause the dots to appear denser, the further away they were viewed.

The situation can also be visualised by thinking of circles drawn on the various geometries. If a circle were to be drawn on a sphere, its circumference would be smaller than $2\pi r$. Thus, if it were to be mapped onto a flat surface, the circumference would have to be stretched. Conversely, a circle drawn on a hyperboloid would have a circumference greater than $2\pi r$, and the circumference would need to be compressed in order to map it on a flat surface.

If the rubber sheet is replaced with the spacetime continuum, and the painted dots with galaxies, then it can be seen how easily this analogy relates to our Universe. In principle, it should be possible for the astronomer to calculate the curvature of the Universe from source count observations. However, a number of complicating factors renders this more difficult than would appear. As already stated, the expansion of the Universe causes us to observe an increase in density. We must try to separate the two phenomena to determine whether there is any residual difference in density. The second problem is to ensure that all of the galaxies in the sample are detected. Looking further and further into space, the fainter members become much more difficult to observe. Therefore, it is very easy to observe an over-density of galaxies in the nearby Universe because the distant galaxies simply cannot be seen. These various challenges have so far conspired to render the source count observations inconclusive.

The second possibility for determining whether or not spacetime is flat is to utilise our concept of the astronomical census, and thus calculate the amount of mass in the Universe. It was asserted in the previous chapter that a component of the Universe is in the form of non-baryonic matter. It is now time to consider a more definitive figure to determine just how much of the Universe is composed of exotic particles. To begin with, we must determine just how much of it is baryonic. Estimating the amount of luminous baryonic matter is relatively straightforward, and although it is subject to a fair margin of error, the result is very important. The density of luminous baryonic matter, ρ_{lb} , is found to be severely lacking compared with what would be necessary to close the Universe:

$$\Omega_{lb} = \frac{\rho_{lb}}{\rho_{crit}} \approx 0.01 - 0.03 \qquad (11.3)$$

The discrepancy between the estimates of the density of luminous matter and the critical density has become known as the 'missing mass' problem. Our first step towards solving this problem lies in our attempts to estimate just how much baryonic matter there is in total, because almost certainly there are vast amounts which are in non-luminous forms: for example, in interstellar clouds in the haloes of galaxies. Mass determinations from the rotation curves of galaxies show that

$$\frac{M_{total}}{M_{lu\,min\,ous}} \approx 2 - 10 \qquad (11.4)$$

The majority of this non-luminous matter must be contained within the halo in order to produce the shape of the rotation curves. Very high-contrast image processing has revealed the haloes of several galaxies. These images have shown that haloes are often flattened systems which are oblate in shape. This single observation bolsters the idea that they are predominantly baryonic, because as orbiting baryonic matter loses energy it will tend to flatten into a disc.

The objects comprising the baryonic component of galactic halos are known generically as MACHOs – massive and compact halo objects. It is thought that they are mostly stellar remnants such as neutron stars and black holes, or failed stars known as brown dwarfs. The best method for directly detecting these MACHOs is to watch for their effect on distant, background stars. If the MACHO's orbit carries it in front of a star, the star will be gravitationally lensed. The gravity of the MACHO is only weak, however, and the images it causes will not be distinguishable. Instead they will combine to make it appear as if the background star has brightened. These MACHO events, as they are called, have been observed several times – most notably in 1993, when a star in the Large Magellanic Cloud was observed to brighten in response to the passage of an unseen object within the halo of our Galaxy. Thus, we can produce an approximate figure for the total density of baryonic matter, ρ_b, in both luminous and non-luminous forms:

$$\Omega_b = \frac{\rho_b}{\rho_{crit}} \approx 0.1 \qquad (11.5)$$

Corroborating this picture is the present-day abundance of the hydrogen isotope deuterium. During the early Universe, this was formed as an intermediate part of the proton–proton chain which results in helium. This reaction is represented in equation (1.1), with the deuterium being produced in the first stage. Thus, the abundance of deuterium can be seen as a measure of how efficiently the helium fusion process took place in the era of nucleosynthesis. If it were 100% efficient, then there should be no deuterium in today's Universe. This is because this isotope is not synthesized in any great quantities by the present-day Universe. If it is part of a star, deuterium is destroyed. Young stars – such a T Tauri stars – can be spectroscopically observed to contain deuterium, but older stars show no trace of this element.

The estimate of the ratio of deuterium to hydrogen in interstellar environments can thus provide a means of estimating the amount of baryonic matter in the Universe. This is because the density of baryons during the era of nucleosynthesis is one of the defining conditions for its efficiency. Had the density been too great, there would be no deuterium left, because it would all have been converted into helium. With insufficient baryons, the abundance of deuterium would be much higher. Using this idea we can estimate the density of baryons based upon our knowledge of nuclear reactions and the observed quantities of deuterium in today's Universe. From this we conclude that there can be no more than 10% of the critical density in the form of baryonic matter, otherwise the deuterium abundance would be incompatible with current observations. This would seem to imply that the Universe was open. However, if we ascend the size scale to clusters of galaxies, and estimate the amount of luminous matter as a ratio to the total mass given by dynamical observations, we find that

$$\frac{M_{total}}{M_{lu\,min\,ous}} \approx 10 - 50 \tag{11.6}$$

Thus we have clear evidence that the Universe contains more matter than its baryonic content. Continuing up the size scale, the largest-scale streaming motions indicate that

$$\frac{M_{total}}{M_{lu\,min\,ous}} \approx 50 - 100 \tag{11.7}$$

This ratio provides, at its upper limit, enough matter to close the Universe, because in equation (11.3) we stated that $\Omega_{lb} \sim 0.01$. If we live in an inflationary Universe, then the unmistakable conclusion is that non-baryonic dark matter may provide up to 99% of the matter in our Universe! Our estimate of the amount of baryonic dark matter led us to equation (11.5) and an estimate of $\Omega_b \sim 0.1$, but that still means that 90% of the Universe is in exotic forms of matter which we may not yet have even discovered. This is quite sobering for anyone who is tempted into thinking that all that remains is to dot the 'i's and cross the 't's!

If the majority of the dark matter is cold dark matter – which from observational and modelling constraints seems probable – then there is an interesting point to note about the scale on which the cosmological principle is applicable. In Chapter 4 it was stated that the Universe is homogeneous only on scales greater than 300–600 million light years. That estimate was based upon a redshift survey which considered only the luminous baryonic matter. If this type of matter represents only 1% of the total density of the Universe, then can we really assume that it traces the real mass distribution? Luminous baryonic matter may not be as efficient as we had hoped at tracing out the matter content of the Universe. If the density of dark matter followed the density of luminous matter, than we should not see an increase in the ratio of

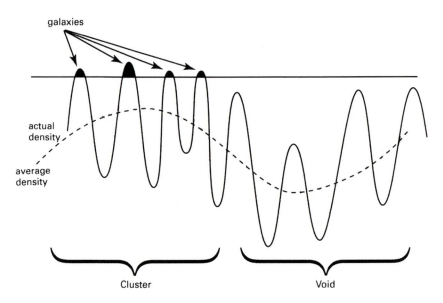

Figure 11.1. In the biased cold dark matter models of galaxy formation, only those conglomerations of matter which cross a certain mass limit will form bright galaxies. These types of model are in reasonably good agreement with observations.

luminous to total mass as we increase the scale from individual galaxies to clusters to superclusters. If, however, the dark matter is spread fairly uniformly throughout space, then, as the volume under investigation increases, so the mass ratio will also increase.

Hot dark matter would naturally behave in this way because, travelling at high speeds, it would resist clumping. Cold dark matter can also behave in this way, and this introduces us to the concept of bias in our models. Biasing in a cold dark matter Universe implies that not all concentrations of density will become galaxies. If we imagine that large-scale undulations are present in the cold dark matter, representing superclusters and voids, then, superimposed on these are fluctuations in the baryonic matter. Only when the total density of both types of matter exceeds a certain level does a galaxy form. This allows for a much more homogeneous distribution of matter throughout the Universe than is indicated from the simple distribution of luminous galaxies (see Figure 11.1).

11.6   COSMOLOGICAL MODELS

The Friedmann equation (4.29) can be used to construct graphs of the scale factor versus time, which allows a visual interpretation of various universal geometries. In the first and simplest case, we shall examine an open Universe which is filled with nothing but radiation. This would have been the case with our Universe if the C, P and T invariances had all held individually and precisely. The various models are named after the people who first devised them, and this particular model is known as the Milne model. We shall begin with the Friedmann equation (5.29), but substitute for the Hubble constant using equation (4.31). After a trivial simplification, we obtain the Friedmann equation in the following form:

$$\dot{R}^2 = \frac{8\pi G\rho R^2}{3} - kc^2 \tag{11.8}$$

In a strict sense, our constant k is not the same as the k in equation (5.29), because we have changed its sign convention to retain the energy constant definitions of the open and closed Universes. We have also divided it by c^2 in order to give it the appearance of a general relativistic interpretation. In an open radiation-filled Universe, $\rho = 0$ and $k = -1$. Substitution of these values leads to

$$\dot{R} = \pm c \tag{11.9}$$

This equation reveals that the rate of change of the scale factor is a constant; that is, the rate is neither accelerating nor decelerating. The constant in this case happens to be the speed of light, and equation (11.9) can be integrated to produce the explicit time-dependent evolution of the scale factor:

$$R = \pm ct \tag{11.10}$$

In a radiation-only open Universe, the rate of cosmic expansion is a constant with a magnitude of the speed of light. Note that the T invariance is present in this model due to the inclusion of the $\pm$ sign (caused by the square root of c^2). This means that a radiation-filled Universe could evolve in exactly the same way, regardless of whether time is running forwards or backwards (see Figure 11.2).

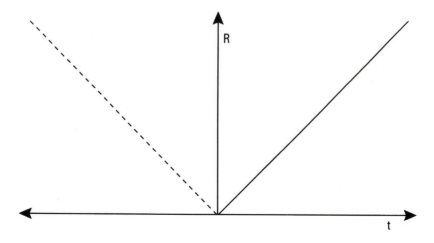

Figure 11.2. The Milne cosmological model provides an idea of how our Universe would have expanded had it contained only radiation. In this model, the Hubble time is the precise age of the Universe.

The age of a radiation-filled Universe can be easily derived. In Chapter 6 the notion of the Hubble time was mentioned. The Hubble time, τ, is the reciprocal of the Hubble constant:

$$\tau = \frac{1}{H_0} \tag{11.11}$$

The Hubble time can be thought of as the time it would take for a celestial object to double its distance if that object is not subjected to any acceleration or deceleration. This would imply that it is also the time in which it has taken the object to reach its current distance. Thus, the Hubble time is an estimate of the age of the Universe if the expansion rate remains constant. In a radiation-dominated Universe, equation (11.9) reveals that the expansion rate does indeed remain constant.

Combining equations (5.31) and (11.11) allows us to calculate the Hubble time in the following way:

$$\tau = \frac{R}{\dot{R}} \tag{11.12}$$

Substituting equations (11.9) and (11.10) into equation (11.12) produces

$$\tau = t \tag{11.13}$$

Thus, in the case of the Milne model, the Hubble time produces an accurate figure for the age of the Universe.

The next case we wish to consider is that of a Universe which contains the critical density of matter. In this case, $\rho = \rho_c > 0$ and $k = 0$. It is included here because it is mathematically the next simplest case; but it is also highly instructive, because these are the parameters of an inflationary Universe. It is known as the Einstein–de Sitter model and, once again, some simplification of the standard Friedmann equation is necessary. In this case, an actual value must be substituted, because we are dealing with a non-zero density of matter. It was stated in Chapter 5 that the density of matter alters with the inverse cube of the scale factor. This allows us to construct an equation for the density, ρ, of the Universe at any time, from its value at the present epoch, ρ_0, and the ratio of the present scale factor R_0, to the scale factor at any time, R:

$$\rho = \rho_0 \left(\frac{R_0}{R}\right)^3 \tag{11.14}$$

When this equation and the energy constant are substituted into the Friedmann equation, we obtain

$$\dot{R}^2 = \frac{8\pi G \rho_0 R_0^3}{3R} \tag{11.15}$$

From this, we can recognise the origin of equation (5.32), because we can define a constant, A:

$$A = \frac{8\pi G\rho_0 R_0{}^3}{3} \tag{11.16}$$

and then state the proportionality:

$$\dot{R}^2 = \left(\frac{dR}{dt}\right)^2 \alpha \frac{1}{R} \tag{11.17}$$

which is equivalent to equation (5.32). Equation (11.17) can then be arranged into the integral:

$$\int_0^R R^{1/2} dR \; \alpha \int_0^R dt \tag{11.18}$$

which evaluates to equation (5.30) and provides us with the time dependency of the scale factor:

$$R = at^{2/3} \tag{11.19}$$

This time the scale factor of the Universe is again dependent on a constant, a, but this is not the speed of light, and the scale factor alters with time because of the term involving t. Differentiation determines the rate of change of the scale factor with time:

$$\dot{R} = \frac{2}{3} at^{-1/3} \tag{11.20}$$

The age of an Einstein–de Sitter Universe is expressed by

$$\tau = \frac{R}{\dot{R}} = \frac{3}{2} \frac{at^{2/3}}{at^{-1/3}} = \frac{3}{2} t \tag{11.21}$$

which states that the Hubble time is an overestimate of the age by a factor of two-thirds (see Figure 11.3). This means that, in an Einstein–de Sitter Universe, a Hubble constant of 100km/s/Mpc would correspond to an age of just 6,600 million years! This is far too short a period of time for our Universe to have been in existence, based upon estimate of the age of some stars. The members of some globular clusters, for instance, are approximately 10,000 million years old. We shall return to this dilemma when we consider the cosmological constant.

Open and closed Universes which contain matter are more complex using the Friedmann equation, but they can be reduced to diagrams showing the way in which the scale factor changes with time (see Figure 11.4).

Because the rate of change of the scale factor included in equation (11.20) is a function of time, it too changes with time. In other words, the universal expansion can either accelerate or decelerate. We know (from our consideration of the fundamental forces of nature in Chapter 3) that gravity is the shaping force of the Universe on the largest scales, and so we would expect a deceleration rather than an acceleration. If we differentiate equation (11.15), the deceleration equation is

$$\ddot{R} = -\frac{4\pi G\rho R}{3} \tag{11.22}$$

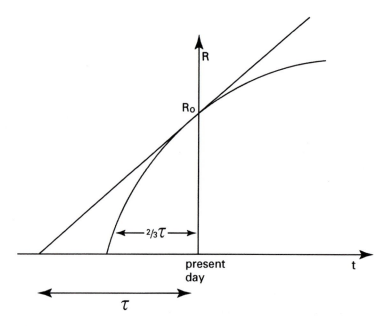

Figure 11.3. The Einstein–de Sitter cosmological model shows how a 'flat' universe expands because it contains both matter and radiation. In this case, gravity gradually slows the expansion and so the Hubble time is an overestimate of the age of the Universe.

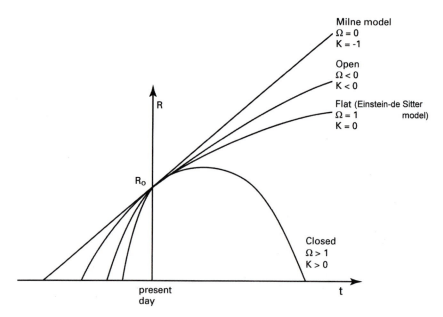

Figure 11.4. Open and closed models of the Universe differ from the 'flat' Universe in that a closed Universe will eventually collapse and an open Universe will expand forever.

The deceleration is usually enveloped in a dimensionless quantity known as the deceleration parameter, q, which is defined in the following way:

$$q = - \frac{R\ddot{R}}{\dot{R}^2} \qquad (11.23)$$

In the case of the Milne model, $q = 0$; in a 'flat' Einstein–de Sitter Universe, $q = 0.5$; open Universes are those in which $q < 0.5$; and in closed Universes, $q > 0.5$.

11.7   GENERAL RELATIVITY

So far, everything in this book has been discussed within the bounds of Newtonian physics. General relativity has been referred to several times, and yet until now the mathematics has excluded it! General relativity is inextricably linked to cosmology, however, and is unavoidable in its continued study. The reason for this is that the Universe is governed, on its largest scales, by the gravity of the objects it contains. Its entire evolution is determined by the amount of mass within it, and the only applicable gravitational theory that we possess is general relativity.

The theory of general relatively evolved from Einstein's theory of special relativity, in which he investigated the effects of travelling at speeds close to that of light. The special theory was restricted because it dealt only with unaccelerated (inertial) cases of motion, and Einstein wished to extend his work to encompass accelerated frames of motion. It has been known since the time of Galileo – from his legendary (but unoriginal) experiment of dropping objects with different masses from the leaning tower of Pisa – that the Earth's gravitational field acts equally on all objects, regardless of their individual mass. The action of the Earth's gravity is to induce objects to accelerate, which is why general relativity can be used to investigate gravity.

The statement that the Earth's gravitational field acts on all objects equally is an approximation. The force of gravity is expressed by equation (1.3), in which, if m_1 refers to the mass of the Earth, it can be seen that the strength of the gravitational force is directly proportional to the mass of the object, m_2, upon which it acts, multiplied by m_1. In the case of the Earth, all droppable objects have a negligible mass in comparison to our planet, and so the acceleration produced by the force appears to be a constant, regardless of the nature of the falling object. This proportionality is also applicable to some other forces – notably, centrifugal force and Coriolis force. It allows us to define a set of forces known as kinematic forces, which are all indistinguishable from accelerations and directly proportional to the mass of the objects upon which they act.

The mass, m_2, in equation (1.3) is called the gravitational mass, m_g, because it causes the force of gravity. Newton's second law of motion states that the inertial mass of an object is the quotient of the force, F, acting upon it to the acceleration, a, which that force is producing:

$$m_i = \frac{F}{a} \qquad\qquad (11.24)$$

The principle of equivalence is the concise statement of the experimental fact that inertial mass and the gravitational mass are exactly the same to a very large degree of accuracy. Using the principle of equivalence we can easily show that all objects – regardless of their mass – are accelerated by a gravitational field in the same way, by simply combining equations (1.3) and (11.24), which then simplifies to

$$a = \frac{Gm_1}{r^2} \qquad\qquad (11.25)$$

Einstein encapsulated his principle of equivalence in a series of four 'thought experiments' which sought to describe what an observer inside a sealed 'box' (less sadistically termed a laboratory) could or could not say about his location and state of motion.

Imagine, for example, a lift containing a person and a ball. The lift is sealed in such a way that no observation of the outside world is possible. In the first experiment the lift is taken into space – well away from any planet or star, which would exert a gravitational pull upon it. A rocket motor is attached to the base of the lift, and the whole assembly is accelerated to 9.8 m/s^2 by ignition of the rocket. The person inside the lift then releases the ball and watches as it falls to the floor. The acceleration of the ball towards the floor would be equal to – but in the opposite direction from – the acceleration of the lift.

In the second experiment, the rocket motors are turned off. The lift would continue to drift at its final velocity. This time, if the ball were to be picked up and released, it would hang motionlessly in mid-air, because no forces would be acting on the lift. The occupant would also feel weightless.

In the third experiment, the lift is brought to Earth and suspended, unmoving, in a lift shaft. The ball is released, and we know from common sense that it would accelerate towards the floor. The magnitude of the acceleration caused by the Earth's gravity is 9.8 m/s^2.

In the fourth and final experiment, the lift is cut from its suspension and allowed to fall freely down the shaft. The lift and its contents (the occupant and the ball) are all accelerated under gravity so that this time, when the ball is released it actually appears to remain stationary inside the lift, since it is already falling at 9.8 m/s^2. Again, the occupant feels weightless (see Figure 11.5).

To the person inside the lift, the first and third experiments are indistinguishable, as are the second and fourth. In the first experiment the acceleration on the ball is caused by the force of the rocket. This is an ordinary example of Newton's second law of motion, and so the mass involved is the inertial mass. In the third experiment the force is gravitational in origin, and concerns itself with the gravitational mass. Both, however, cause the ball to accelerate downwards and, because of the principle of equivalence, are indistinguishable from one another. Imagine further what would happen to a rocket accelerating away from the Earth at 9.8 m/s^2. Since the

Figure 11.5. The principle of equivalence. Einstein constructed four 'thought experiments' to demonstrate that gravity was indistinguishable from being inside an accelerating frame of reference. In this way, his general theory of relativity could be used to explain the actions of a gravitational field, even though it had been designed to understand the consequences of accelerated motion. (The details of the four experiments are discussed in the main text.)

downward pull of gravity would tend to accelerate it in the opposite direction at exactly the same rate, the rocket would not move.

The conclusions of these 'thought experiments' can be expressed more concisely. Einstein stated that accelerations and gravitational fields are indistinguishable from one another, so that a correctly chosen and applied acceleration can counteract the effects of a gravitational field, and vice versa. In essence, this is the principle of equivalence, which demonstrates why general relativity – which sought to describe an accelerating frame of reference's view of the Universe – also describes gravity.

The next concept which is central to general relativity is the spacetime continuum. In 1908, Hermann Minkowski first propounded the idea of a four-dimensional representation of the Universe which described events by referencing their three-dimensional spatial co-ordinates (x,y,z) and the time, t, at which they took place.

In three dimensions, the distance between two points in a Cartesian grid is represented

$$\Delta s^2 = \Delta x^2 + \Delta y^2 + \Delta z^2 \tag{11.26}$$

In spherical polar co-ordinates, this is written as

$$\Delta s^2 = \Delta r^2 + r^2 (\Delta \phi^2 + \sin^2 \phi \Delta \theta^2) \tag{11.27}$$

To extend this idea to a spacetime framework we can imagine two events, each one characterised by three spatial co-ordinates and a time. To transform the times into units of distance, both must be multiplied by the speed of light. The interval between the two can then be calculated:

$$\Delta s^2 = \Delta t^2 + \frac{1}{c^2} (\Delta x^2 + \Delta y^2 + \Delta z^2) \tag{11.28}$$

In spherical polar co-ordinates this becomes

$$\Delta s^2 = \Delta t^2 + \frac{1}{c^2} (\Delta r^2 + r^2 \Delta \phi^2 + r^2 \sin^2 \phi \Delta \theta^2) \tag{11.29}$$

The concept of space-like and time-like separation (introduced in Chapter 4) can now be treated mathematically. The interval between two events is space-like if an observer cannot be present at both events, and time-like if the observer can be present at both events. In the case of a space-like separation, $\Delta s^2 < 0$, while time-like separation is characterised by $\Delta s^2 > 0$. The dividing line between these two possibilities is represented by the conical surface of the light cone in Figure 4.3. These are the paths followed by light rays from the first event. If the second event lies on the edge of the first event's light cone – that is, only rays of light leaving the first event can be present at the second event – then the interval between them is null, because $\Delta s^2 = 0$. Each time an event – such as a supernova – is observed, the interval between the event and the act of observing it is null.

In the same way that Δs^2 of equation (11.27) describes a tiny portion of space between two points, so the Δs^2 of equation (11.29) describes a tiny portion of spacetime between two events. Because of the cosmological principle – which states that the Universe is both homogeneous and isotropic – Δs^2 must be the same at each point on the spacetime continuum. This severely restricts the possibilities, and leads to the positive and negative curvatures which have been considered. The 'flat' case of no curvature is also allowed by the cosmological principle.

Equation (11.29) is known as the Minkowski metric. A metric is the term for an equation which can mathematically describe a four-dimensional structure such as the spacetime continuum. In order to take into account the cosmological principle, Robertson and Walker modified it to become:

$$\Delta s^2 \;=\; \Delta t^2 - \frac{R^2}{c^2}\left(\frac{\Delta\sigma^2}{1-k\sigma^2} + \sigma^2\,\Delta\phi^2 + \sigma^2\sin^2\phi\Delta\theta^2\right) \tag{11.30}$$

where $\sigma = r/R$. In this equation, σ is a co-moving radial ordinate, and k is the energy encountered in the Friedmann equation. In this general relativistic context, it is usually known as the curvature constant.

To illustrate that this metric describes the spacetime continuum, let us consider one of its fundamental qualities: expansion. Equation (9.1) expresses the spectral ratio, which we stated (without proof) determined the expansion factor of the Universe. If the redshifting of radiation is indeed produced by the expansion of the spacetime continuum, then by using the Robertson–Walker metric, we should be able to determine this behaviour, because the expansion is embodied in the scale factor, R, in equation (11.30).

As defined earlier in this chapter, the observation of a distant galaxy is said to be a null interval, in which $\Delta s^2 = 0$. The galaxy is at $\sigma = \sigma_e$, and the time of emission of the first wave crest is t_e. The second wave crest is then released at $t_e + \Delta t_e$. The time of observation, t_o, is the point at which the observer, at $\sigma = 0$, sees the first wave crest. The second wave crest is then observed at $t_e + \Delta t_e$. Galaxies will not exhibit any appreciable change in their positions on the sky plane because of their great distances, and so $d\theta = d\phi = 0$. Using this to simplify the Robertson–Walker metric we obtain

$$0 \;=\; \Delta t^2 - \frac{R^2}{c^2}\frac{\Delta\sigma^2}{1-k\sigma^2} \tag{11.31}$$

This can be rearranged and integrated to derive the change in the scale factor between the time of emission and observation of the first crest:

$$\int_{t_e}^{t_o}\frac{dt}{R(t)} = \frac{1}{c}\int_{\sigma_e}^{0}\frac{d\sigma}{\sqrt{(1-k\sigma^2)}} \tag{11.32}$$

A similar integral can be formulated for the emission and observation of the second wave crest. This new equation would have a right-hand side identical to equation (11.32), because the limits will still be 0 and σ_e (remembering that σ is a co-moving radial ordinate). Thus, equation (11.32) transforms into the following equality:

$$\int_{t_e+\Delta t_e}^{t_o+\Delta t_o}\frac{dt}{R(t)} = \int_{t_e}^{t_o}\frac{dt}{R(t)} \tag{11.33}$$

By investigation of the limits, equation (11.33) can be substituted for

$$\int_{t_o}^{t_o+\Delta t_o}\frac{dt}{R(t)} = \int_{t_e}^{t_e+\Delta t_e}\frac{dt_e}{R(t)} \tag{11.34}$$

The time between the emission of the two wave crests is very small; and the time between the observation of them is also very small. Over very small intervals the scale factor R(t) can be regarded as constant. Therefore:

$$\frac{dt_o}{R(t_o)} = \frac{dt_e}{R(t_e)} \qquad (11.35)$$

If the slight modification of equation (2.1), $\lambda = cdt$, is incorporated, then equation (11.35) can be written as

$$\frac{\lambda_o}{\lambda_e} = \frac{R_o}{R_e} \qquad (11.36)$$

which is equivalent to the spectral ratio, $1+z$, in equation (9.1). Thus, our earlier assertion – that the spectral ratio is a measure of the expansion factor of the Universe – is shown here to be true. This is because we have proved that the ratio of the observed and emitted wavelengths is equal to the ratio of the scale factors at the epochs of observation and emission.

11.8   THE COSMOLOGICAL CONSTANT

Before the discovery of the expanding Universe, nearly all cosmologists assumed that the cosmos was static. The only two who suspected otherwise were Aleksandr Friedmann and George Lemaître (as discussed in Chapter 5). They were both early exponents of Einstein's general relativity. Chapter 1 explained how Einstein had initially been perplexed at his theory's failure to predict a static Universe. Finally, his consternation led him to introduce a new term which he called L, the cosmological constant. It is often stated (especially by those wishing to deride Einstein's work) that this was a 'fiddle factor' designed to modify the equations, although this is probably a little harsh, because in the theory there is some justification for its existence.

When Einstein set this term to a positive number it provided a cosmic repulsion which could be fine-tuned to resist the gravitational attraction and hold the Universe static. The modified form of the Friedmann equation is

$$\dot{R}^2 = \frac{8\pi G\rho R^2}{3} - kc^2 + \frac{\Lambda R^2}{3} \qquad (11.37)$$

and the deceleration equation becomes

$$\ddot{R} = -\frac{4\pi G\rho R}{3} + \frac{\Lambda R}{3} \qquad (11.38)$$

To determine the value of Λ, which holds the Universe static, Einstein placed both equations (11.37) and (11.38) equal to 0, $k = +1$ and solved them simultaneously. This critical value of $\Lambda - \Lambda_c$ – was shown by Hubble to be superfluous when he proved the expanding Universe, and Einstein immediately referred to his constant as his greatest mistake, and banished it from further consideration. However, this has not prevented its occasional use by cosmologists, as it can be used to provide an 'escape route' from the age dilemma pointed out earlier. George Lemaître postulated

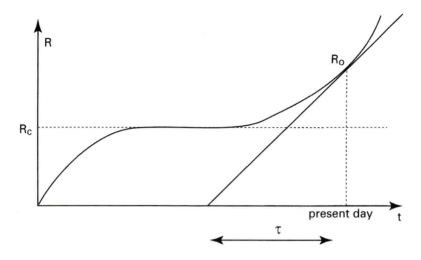

Figure 11.6. A Universe containing a cosmological constant can be a Lemaître hesitation Universe. In this case, the Hubble time is an underestimate of the age of the Universe.

a Universe in which the value of the cosmological constant was very slightly higher than the critical value:

$$\Lambda = \Lambda_c (1 + \varepsilon) \tag{11.39}$$

where $\varepsilon \ll 1$. This causes a 'hesitation' Universe in which the expansion is severely arrested for a cosmological epoch before continuing. This causes estimates of the Hubble time to be too small when compared with the actual age of the Universe (see Figure 11.6). If the period of arrested expansion occurs at a scale factor, R_c, then an observational consequence of this type of Universe is that there will be a concentration of objects at a redshift, z, defined as

$$1 + z = \frac{R_0}{R_c} \tag{11.40}$$

Although there has never been conclusive observational evidence to support this theory, an oddity worth noting is the peak in the space density of quasars. This observational proof that quasars appear to cluster around redshift 2 was mentioned briefly at the end of Chapter 7. Could this be an indication that there was considerable slowing in the expansion of the Universe when it was just one-third of its present size? Although for unconventional astronomers the idea is tantalising, the weight of evidence instead suggests that it is caused by an evolutionary effect, with quasars simply reaching a vigorous state of activity by that stage in the evolution of the Universe. Deep surveys of Seyfert galaxies show no such clustering at $z = 2$, and so the 'hesitation' Universe theory must now be regarded as defunct.

The same could once have been said about the cosmological constant, but in recent years it has again been considered by cosmologists. This is because its interpretation is based on the energy of space – a vacuum energy. The idea derives from particle physics, which considers that space is not empty but seethes with virtual particles. Under the right conditions, the energy necessary to create these virtual particles provides a buoyancy which resists the collapse of the Universe that gravity seeks to cause. Thus, as well as prolonging the life of the Universe, this natural buoyancy, or 'springiness', would also provide a support mechanism, which would lead to the formation of very large-scale structures. The ultimate reason for the newly-found excitement concerning Λ, is that by endowing the vacuum with energy, an additional gravitational effect is created, because energy is equivalent to mass (equation (1.2)). This implies that a Universe with a cosmological constant may not need any dark matter, because it is the energy which provides the additional gravity to hold superclusters, and so on, together. The cosmological constant is therefore a seductive concept. If the inflationary theory is correct and we *do* live in a flat Universe, a cosmological constant would modify the condition for 'flatness' in the following way:

$$\Omega + \Lambda = 1 \tag{11.41}$$

Another observational test of the cosmological constant is that gravitational lenses should be more numerous in a Universe with a non-zero Λ. The higher the magnitude of the cosmological constant, the greater the number of lens systems. Estimates of Ω suggest that Λ might be as high as 0.8 or 0.9. From theory, this should result in an increase in the number of gravitational lenses by a factor of 10–100. Unfortunately, this is not observed. A recent Hubble Space Telescope survey – analysed by John Bahcall, of the Institute for Advanced Study in Princeton – included almost 500 quasars and found only a few of lensed systems. Gravitational lenses are so scarce that some scientists are not convinced that this is enough to completely rule out Λ. However, another series of observations may clarify Λ.

11.9   THE ACCELERATING UNIVERSE THEORY

What if the cosmological constant produced more than just buoyancy? If this were the case, it could actually help to drive the expansion of the Universe and force it to accelerate. This is exactly what is believed to have been found in the study of Type 1a supernovae by the Supernova Cosmology Project (discussed in Chapter 6).

The supernovae studied were about 15% dimmer than they should be if $\Lambda = 0$. Both teams were therefore forced to conclude that $\Lambda > 0$ and was currently causing the expansion of the Universe to accelerate. A positive cosmological constant therefore manifests itself as some kind of antigravity force (see Figure 11.7).

Chapter 3 discussed the creation of gauge bosons (by borrowing energy from surrounding space) and it has already been mentioned that a seething mass of virtual particles is expected in the quantum mechanical view of the vacuum. There is even proof of these virtual particles. Equation (3.3) used the Heisenberg uncertainty principle to link the quantities of energy and time, and to show that the more energy borrowed, the less time the virtual particle could exist.

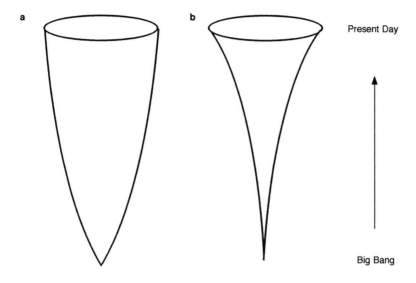

Figure 11.7. Accelerating Universes. Instead of the traditional view (a) in which the expansion rate of the Universe decreases with time, a large enough cosmological constant (b) would cause the Universe's expansion rate to accelerate.

The consideration of wave–particle duality (discussed in Chapter 3) revealed that the energy contained in a particle could manifest itself as a wave motion. Different virtual particle energies would correspond to different wavelengths: the more energy, the shorter the wavelengths.

If two metal plates were to be erected parallel to each other with a tiny gap between them, then virtual particles would be expected to be spontaneously forming around them and in the gap between them. In the gap, however, not all wavelengths of virtual particle would be capable of forming, because there would be insufficient physical space for them to exist. Inside the plates, certain wavelengths – and therefore energies of virtual particles – would be inhibited from forming. Outside the plates, no such restrictions apply, and virtual particles of all wavelengths could form. There would therefore be less energy inside the plates than outside, which would push them together. This is the Casimir effect, which has been measured in the laboratory. So, the production of virtual particles has been experimentally verified. The only place where they could be obtaining their energy is the vacuum energy of the Universe, as represented by a positive cosmological constant. It was stated in Chapter 5 (in the section discussing inflationary cosmology) that vacuum energy drives the expansion of the cosmos. The existence of virtual particles implies that the expansion of our Universe will slowly accelerate, with time creating an accelerating Universe.

The first corroborating evidence for this view derives from observations of supernovae at large cosmological distances, obtained by the Supernova Cosmology Project. Even so, some astronomers still feel that the results are a little tentative, and that the obscuring effects of dust grains could account for the faintness of the supernovae. Over the next few years there will doubtless be considerable activity to confirm or deny this intriguing possibility. Regardless of whether or not this work is corroborated, the cosmological constant will always remain an important – if somewhat eccentric – cosmological parameter.

11.10 THE WAY FORWARD

In some ways, cosmological theories have developed far faster than have observational methods. Only when fundamental new observations are procured can cosmologists really test their current theories. For example: when COBE discovered the anisotropies in the microwave background radiation, the upper limits they set by them – in terms of the strength of the fluctuations and their angular extent – discounted some of the more outlandish cosmological possibilities. Only the models which are discussed in this book have survived. New observations are continually required, allowing theoretical predictions to be compared with actual data. With this in mind, the final section of this book previews some of the new technologies and observational techniques which will enhance our understanding of the Universe and allow cosmology to develop.

Without doubt, two of the most important cosmological images are the Hubble Deep Fields. The Hubble Space Telescope is also involved in another major cosmological investigation. A stated aim of the science team responsible for the orbiting telescope is that it should try to refine the Hubble constant to within 10% of its actual value. This was mentioned in Chapter 6, but warrants reiteration. So far, the available Hubble data would appear to be closing in on a figure consistent with the average value of 75 km/s/Mpc that we have used throughout this book.

The successor to the Hubble Space Telescope – the Next Generation Space Telescope (NGST) – is now at planning stage, and is scheduled for launch in 2007 (see Figure 11.8). It will be much larger than the HST, with preliminary plans calling for an 8-m mirror that is optimised to carry out observations in the infrared. It can therefore be used more effectively to study extrasolar planets, star formation and the high-redshift Universe.

At present, one of the most exciting projects in observational cosmology – the 2-degree Field (2dF) mapping project – is being undertaken by a joint British and Australian team at the Anglo-Australian Observatory at Siding Spring, New South Wales. Conventional telescopes are limited to obtaining the spectrum of a single galaxy during one observation, but the 2dF will simultaneously record up to 400.

The 2dF instrument is self-contained, and is mounted on the 150-inch (3.9–m) Anglo-Australian Telescope. A system of lenses converts the telescope into a wide-angle instrument with a field of $2°$. A robot then places each of the 400 optical fibres precisely over the galaxy images in the field of view, and their light is simultaneously fed to a spectrograph, which records the information. These spectra can be converted into redshifts, and equation (1.4) can be used to convert them into actual distances. The ability to simultaneously record 400 spectra has created much excitement,

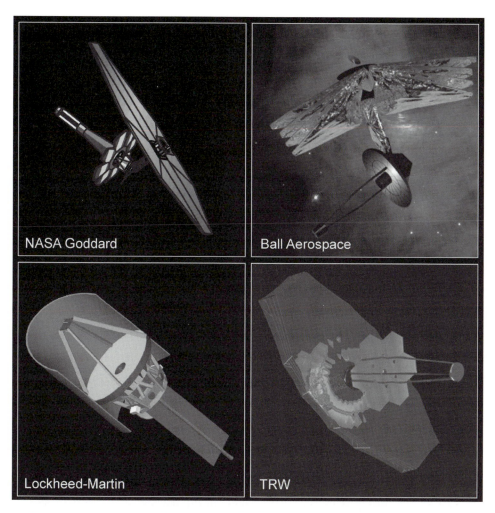

Figure 11.8. Four possible designs for the Next Generation Space Telescope – the successor to the Hubble Space Telescope. It will possess a large-diameter mirror and be optimised to work at infrared wavelengths. (Photograph reproduced courtesy of NASA.)

especially among those working on the instrument at the AAO. They feel confident that sufficient galaxy redshifts can be obtained within two years of the instrument's operation to produce an accurate and representative view of the three-dimensional structure of the Universe. The promise of the 2dF project has enormous implications for cosmology; in the same way, for example, as a geographer who has been studying the world using a street map of London is suddenly presented with a global atlas!

2dF is now fully operational, and a second redshift mapping project is also underway. This complementary project is an American endeavour – the SLOAN Digital Sky Survey. Whereas 2dF is mapping galaxies in the southern hemisphere,

SLOAN is concentrating on those in the north. Unlike 2dF – which must share observing time – SLOAN is a dedicated telescope and multiple-fibre spectrograph.

These new redshift maps will effectively be three-dimensional maps of the Universe, and will allow much better comparison between the predictions of theory and the actual Universe. The ideas about biased cold dark matter will, for example, be more readily tested. Large-scale streaming motions will become much more apparent, and this will allow the constraints on the position and mass of dark matter deposits to be much better understood. The precipitous rise in known redshifts will also assist in the quest to determine the value of the Hubble constant. In particular, it will help assess the quality of tertiary distance indicators – those which rely on galaxies of the same Hubble type possessing similar physical properties such as diameter and mass.

Other critically important observations are those which will carry out further study of the cosmic microwave background radiation. Since COBE confirmed the existence of the fluctuations, astronomers have wanted to reinvestigate them. This involves the study of this diffuse radiation by using more sensitive telescopes which possess better angular resolution than did COBE. In America, Europe and Russia, equipment is now being designed for that very purpose.

NASA will launch the Microwave Anisotropy Probe (MAP) in 2000. This will enable the first high-resolution observations of the ripples in the microwave background radiation, and is eagerly anticipated. The ultimate mapping mission, however, is a European space probe. The European Space Agency has investigated a spaceprobe design originally named COBRAS/SAMBA, after a British-led team proposed the Cosmic Background Radiation Anisotropy Satellite (COBRAS) and a French team proposed the Satellite for the Measurement of Background Anisotropies (SAMBA). The combined project is the Planck Surveyor, which will be launched in 2007. Whereas COBE mapped the microwave background to a resolution of 7°, Planck Surveyor will map with a resolution of 5–30 arcminutes, depending upon the wavelength being observed (see Figure 11.9). These two new satellites will determine whether the growth of the large-scale structure of the Universe is the result of random fluctuations in the density of gas at the point of decoupling. This is the assumption underlying much of modern cosmology. If it is shown to be in error, there could be some exciting times ahead!

A common factor concerning these projects is that they will not be placed in Earth orbit in the same way as was COBE. The interference of terrestrial and solar signals made the work of COBE rather more difficult than was desirable – and the Earth and the Sun persistently intervened. This was a particular problem with the Sun because the instruments on board COBE were so sensitive, having been designed to detect the background radiation, and a single glimpse of the Sun would have destroyed them. The next generation of microwave satellites will therefore be placed in orbit as far away from the Earth as possible, to enable greater control in pointing them out of harm's way. They will probably be placed at one of the Lagrangian points, either in front of, or behind, the Earth.

In addition to the various satellite missions to study the background radiation, there are several ground-based telescopes undertaking the same work. The

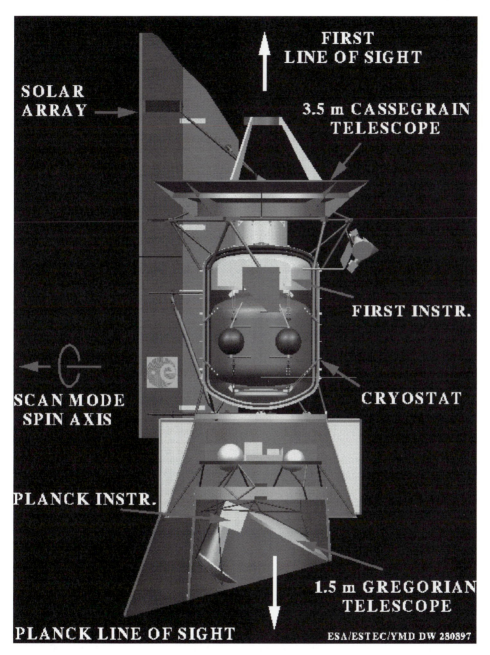

Figure 11.9. The European Space Agency proposes to launch the Planck Surveyor within the next decade, although budget constraints may necessitate its being combined with the Far Infrared and Submillimetre Telescope (FIRST). In this picture, FIRST is at top and Planck is at bottom. (Photograph reproduced courtesy of ESA.)

Cambridge Anisotropy Telescope (CAT) is a three-element interferometer built inside a Second World War bunker to prevent its sensitive detectors from being swamped by terrestrial microwave sources. Whereas COBE found fluctuation at approximately the 10-degree scale, CAT is designed to search for differences on a scale of 10–60 arcminutes.

Another British experiment, located on Tenerife, has already corroborated the COBE results. Other experiments are being set up all over the world. As examples, a series of balloon-borne experiments are being continually undertaken at Princeton University in America, and two other American microwave projects are located at the Center of Astrophysics, located at the geographic South Pole in Antarctica.

Numerous microwave observations will be accrued during the next few years, and these will allow much tighter constraints to be placed upon cosmological theories.

Other instruments which will gradually rise in cosmological significance, as technology progresses, are the neutrino detectors and the gravitational radiation detectors. As mentioned in Chapter 10, both will eventually provide ways of observing the Universe as it was before the decoupling of matter and energy.

As well as technological advances, theoretical advances are required in order to understand the Planck era and, perhaps, even the moment of creation. Formerly, the question of what happened before the Big Bang was usually answered by a (sniffily) delivered reply that time was created in the Big Bang and so the concept of ' before' is meaningless. This has proven unsatisfactory to many scientists, and ideas about the birth of our Universe now abound. Central to the understanding of these ideas is the concept of quantum cosmology – a marriage of quantum theory and general relativity. Mentioned briefly in Chapter 5, this fascinating new branch of cosmology postulates that our Universe was created with ten dimensions, only four of which have survived in order to produce a spacetime continuum. It predicts that the creation of our Universe took place at one specific point in an endless 'multiverse' in which quantum fluctuations appear like bubbles in a pan of boiling water. They are similar to those which, in our Universe, generated the seeds of today's galactic structures. These spontaneous events rapidly expanded and became Universes, some of them similar to our own, but the vast majority of them very different!

Cosmology is a rich field, full of impressive science, technological wonders and concepts far stranger than any science fiction writer could create! The questions of existence have tantalised mankind for so long now that they are an integral part of our everyday lives. At long last, we have made some measure of progress towards explaining how we came to exist. Perhaps we can look forward to understanding even more when the final pieces of the Big Bang theory fall into place. Perhaps we should contemplate the even more exciting possibility that some observations will prove impossible to fit into the Big Bang theory, and that theorists will be left grasping for a totally new idea.

Appendix 1

Physical and astronomical constants

Astronomical unit	AU	1.496×10^{11} m
Boltzmann constant	k	1.38×10^{-23} J/K
Electron volt	eV	1.60×10^{-19} J
Gravitational constant	G	6.67×10^{-11} Nm^2/kg^2
Light year	ly	9.46×10^{15} m
Parsec	pc	3.262 ly
Planck constant	h	6.63×10^{-34} Js
Solar mass	M_{Sun}	1.99×10^{30} kg
Speed of light	c	3.00×10^8 m/s
Stefan–Boltzmann constant	σ	5.67×10^{-8} W/m^2/K^4

Appendix 2

Glossary

Absolute magnitude   The apparent magnitude of a celestial object at a distance of 10 parsecs.
Active galaxy   A galaxy which is emitting large quantities of non-thermal radiation.
Antimatter   The 'mirror image' of matter. Antimatter is opposite in charge but equal in mass and spin to ordinary matter. It is comparatively rare.
Apparent magnitude   The brightness of a celestial object as it appears in the night sky.

Baryon   A particle consisting of three quarks. Baryons are a subset of the hadrons. Neutrons and protons are baryons, as are the heavier, but shorter-lived, particles such as the lambda and the omega.
Big Bang   The initial point of creation.
Billion   For the purposes of cosmology, one billion is defined as an American billion – 1,000 million.
Black hole   A volume of space in which the density of matter is so great that not even light can escape from its gravitational attraction. Inside a black hole, the known laws of physics break down.
Blazar   A highly variable active galaxy which, in general, displays no emission lines in its spectrum.
Blueshift   The increase in the frequency of electromagnetic radiation, caused by the decreasing of the relative distance between the source and observer.
Boson   A particle which does not obey the Pauli exclusion principle. It is denoted by an integer (or zero) spin.
Bottom-up scenario   A galaxy formation scenario in which small galaxies form first. Larger and larger structures are then formed in due course.
Bremsstrahlung radiation   Electromagnetic radiation emitted by electrons interacting with the ions in an ionised gas.

Closed Universe   Any model of the Universe in which the gravity of the matter content can reverse the expansion and cause a collapse.
Cold dark matter   Any dark matter candidate which was non-relativistic at the point of decoupling.

Compton scattering   The scattering of photons by free electrons in an ionised medium.

Compton wavelength   The wavelength of a photon containing the rest energy of a particular particle.

Co-moving co-ordinates   A set of co-ordinates which does not change in an expanding (or otherwise moving) medium; for example the co-ordinates of a distant galaxy do not change just because of the expansion of space.

Copernican principle   States that the Earth possesses neither an ideal nor a privileged position in the Universe.

Cosmic microwave background   A constant flux of electromagnetic radiation which has been redshifted into the microwave region of the spectrum. The photons of cosmic microwave background radiation outnumber the matter particles by 1,000 million to one.

Cosmic substratum   An idealised, smooth cosmic fluid which is spread throughout space evenly, and thus possesses a constant density. It is equal in mass to the constituents of the Universe.

Cosmological principle   States that the Universe is both homogeneous and isotropic to observers at rest within the substratum.

CPT invariance   A symmetry which is believed to hold true for all particles throughout the course of Universal history. It states that matter and antimatter would only react in the same way if the spins of the antimatter particles were reversed and the reaction was caused to run backwards in time.

Dark matter   Any form of matter which exists in a non-luminous form.

Decoupling of matter and energy   A cosmic epoch during which the matter content of the Universe ceased to be ionised. This led to a decrease in the optical depth of the Universe, and the photons of radiation (which we now observe as the cosmic microwave background) were enabled to travel large distances without interacting with matter.

Diameter distance   Any distance to a celestial object which is based upon the use of a standard ruler.

Doppler effect   The alteration in frequency of electromagnetic radiation due to relative motion between the source and observer.

Einstein–de Sitter cosmological model   A Friedmann model of the Universe in which the spacetime continuum is not curved.

Electromagnetism   One of the four fundamental forces of nature. It governs the electric and magnetic interaction between particles.

Electron   A lepton with an electric charge of –1. An electron is also a fermion because it has a spin of one half.

Electroweak force   The combination of the electromagnetic force and the weak nuclear force which takes place at high energy.

Event   A happenstance in the spacetime continuum referenced by three spatial co-ordinates and a complementary temporal ordinate.

Faint blue galaxy   A class of distant, irregularly-shaped galaxy in which a large amount of star formation is taking place.

Fermion   A particle which obeys the Pauli exclusion principle. It is denoted by a half integral spin.

Friedmann equation   A mathematical expression which allows the expansion of the Universe to be studied.

Flatness problem   Poses the question: why, out of an infinite number of possibilities, is our Universe so close to the one special case: the 'flat' Universe?

'Flat' Universe   A universe in which there is no curvature of the spacetime continuum. This means that the kinetic energy of the expansion is exactly balanced by the potential gravitational energy of the matter. Thus, after an infinite amount of time the Universe will stop expanding.

Galaxy   A collection of matter which usually manifests itself by the production of stars.

Galilean transformation   The non-relativistic method of relating observations from one frame of reference to another.

Grand unified theory   Any theory which seeks to unify the strong nuclear force with the electroweak force.

Gravity   One of the four fundamental forces of nature, and the one most different from the other three.

Hadron   Any particle composed of quarks.

Heisenberg uncertainty principle   States that the position and momentum of a particle can be known only to a certain level of precision. The more precisely one quantity is known, the less certain the precision of the other. A similarly linked pair of quantities is the time and energy content in a volume of space.

Helium problem   Poses the question: what physical process caused the current abundance of helium in the Universe?

Horizon problem   Poses the question: why is the cosmic microwave background radiation isotropic across the whole sky, when different regions of the sky could not have been causally connected at the time of decoupling?

Hot dark matter   Any form of dark matter which was relativistic at its point of decoupling.

Hubble classification of galaxies   A morphological classification sequence of galaxies devised by Edwin Hubble. It splits galaxies into ellipticals, lenticulars, spirals, barred spirals and irregulars.

Hubble constant   The constant of proportionality in the Hubble law. Its value must vary with time, so it is often referred to as the Hubble parameter. The Hubble constant is generally used to define the value of the Hubble parameter at the current epoch, and is around 50–100 km/s/Mpc, with possibly a value close to 75 km/s/Mpc.

Hubble flow   The movement of the galaxies away from us, caused by the expansion of the Universe.

Hubble law   The linear proportionality, noticed by Hubble, between the distance of a galaxy and its redshift.

Hubble time   The time it would take for a galaxy to double its distance from the Milky Way. It can therefore also be used to estimate the age of the Universe.

Inflation   A theory which postulates that at 10^{-35} s after the Big Bang the spacetime continuum underwent an intense period of exponential expansion, in response to the separation of the strong nuclear force from the electroweak force. This idea solves the flatness and horizon problems.

Invariant   Any physical property which does not change under the transformation from one frame of reference to another.

IRAS galaxy   Any galaxy which was discovered by the Infrared Astronomical Satellite (IRAS) to possess an excessive amount of infrared emission.

Jeans mass   The critical mass a volume of space must contain before it collapses under the force of its own gravity.

Lepton   A fermion which is not composed of quarks.

Light cone   A cone representing the transmission, at the speed of light, of an event's existence on a spacetime diagram.

Long scale   The cosmological distance scale which uses a Hubble constant of approximately 50 km/s/Mpc.

Lorentz transformation   The transformation which keeps the speed of light invariant between relativistic frames of reference.

Low surface brightness galaxy   A galaxy which is very faint because it contains a very limited number of stars.

Luminosity distance   Any distance to a celestial object which has been calculated using a standard candle.

Lyman-α forest   The appearance of many differentially redshifted Lyman-α absorption lines in a quasar's spectrum, due to intervening hydrogen clouds along our line of sight to the quasar.

MACHO   A massive compact halo object. These are black holes, neutron stars and brown dwarfs, none of which are luminous, and all of which are postulated to exist in the haloes of galaxies. They are a form of dark matter.

Malmquist bias   The systematic distortion in a standard candle's effective range, due to failure in detecting the fainter examples of the standard candle at large distances.

Meson   Any particle composed of two quarks. Examples are the pion (pi-meson) and the kaon.

Milne cosmological model   A Friedmann model of the Universe in which matter does not exist. Only radiation is present in a Milne universe.

Missing mass problem   Poses the question: why does the Universe seem to have much more mass in it than can be seen with a telescope? Dynamical and theoretical constraints place the proportion of missing mass to be around 90–99% of the total mass of the Universe.

Neutron   A baryon composed of one up quark and two down quarks. It possesses no electromagnetic charge, and exists only in atomic nuclei.

Neutrino   A tiny, possibly massless, particle which travels at the speed of light and carries energy and momentum away from interactions involving the weak nuclear force. A possible candidate for hot dark matter.

Nucleosynthesis   The act of building heavier and heavier atomic nuclei from the fusion of protons and other atomic nuclei. The Universe is postulated to have passed through an era of nucleosynthesis lasting about four minutes, during which time helium was produced. This solves the helium problem.

Observer   Anything in receipt of electromagnetic radiation.

Olbers' Paradox   Asks why the night sky is dark. If the Universe were infinite in extent, static, and contained stars (or galaxies) scattered throughout it at random, the night sky should appear as bright as the Sun.

Open Universe   Any model of the Universe which does not contain enough matter to halt its expansion.

Particle accelerator   An experimental device designed to accelerate charged particles to energies comparable with those present during the first minute of the Universe's existence.

Particle horizon   The distance a photon of radiation could have travelled since the creation of the particle.

Pauli exclusion principle   States that particles with half integer spins cannot occupy the same quantum states. This manifests itself as the reason why solid objects cannot exist in the same physical space.

Photon   The fundamental quantum particle of electromagnetic radiation.

Planck era   The first 10^{-43} s of the Universe's existence. Physics can currently explain very little about this time. Quantum gravity is required before quantum cosmology can be fully realised.

Positron   The antimatter counterpart of an electron.

Principle of equivalence   States that inertial mass is indistinguishable from gravitational mass.

Proton   A baryon composed of two up quarks and a down quark. It possesses a positive electromagnetic charge, and exists only in atomic nuclei. A single proton is a hydrogen nucleus.

Quantum cosmology   The study of the Planck era.

Quantum fluctuation   The spontaneous fluctuation of energy in a volume of space. A consequence of the Heisenberg uncertainty principle.

Quantum gravity   A quantum theory of gravity in which gravitons transmit the force between particles, rather than the curvature of the spacetime continuum, as in the general theory of relativity.

Quantum theory   A theory which seeks to explain that the action of forces is a result of the exchange of sub-atomic particles.

Quark   A sub-atomic particle which is a fundamental building block of the hadrons.

Quasar   An intensely bright extragalactic object which superficially resembles a star. Most exist at very high redshifts and are therefore thought to be the nuclei of active galaxies.

Radio galaxy   A galaxy which displays anomalously large radio emission.

Redshift   The decrease in the frequency of electromagnetic radiation frequency, caused by the increasing of the relative distance between the source and observer.

Robertson–Walker metric   An equation which describes the spacetime continuum in a Universe which adheres to the cosmological principle.

Scale factor   An arbitrary measure of the size of the Universe.

Schwarzschild radius   The radius at which a given mass turns into a black hole.

Seyfert galaxy   A spiral galaxy with an overly bright nucleus which is not being produced by stars.

Short scale   The cosmological distance scale which uses a Hubble constant of approximately 100 km/s/Mpc.

Source   Anything which is emitting electromagnetic radiation.

Spacetime continuum   A four-dimensional framework in which events take place.

Sparticles   Hypothetical particles which are predicted by some grand unified theories.

Spectral ratio   The ratio of electromagnetic wavelengths from different cosmic epochs. This determines the expansion factor of the Universe.

Spin   A quantum property of all particles which denotes the intrinsic angular momentum of the particle.

Standard candles   Any luminous celestial object which is more or less constant in its absolute magnitude. It can be used to gauge distances, because the further away it is, the fainter it will appear.

Standard rulers   Any extended celestial object which is more or less constant in its diameter. It can be used to gauge distances, because the further away it is, the smaller it will appear.

Starburst galaxy   A galaxy in which an anomalously large rate of star formation is taking place.

Strong nuclear force   One of the four fundamental forces of nature. It governs the interaction between particles in atomic nuclei.

Sunyaev–Zel'dovich process   Compton scattering between the photons of the cosmic microwave background radiation and electrons in galaxy clusters.

Superforce   The force which is dominant in grand unified theories. It combines the electroweak force with the strong nuclear force.

Symmetry   A set of invariances.

Top-down scenario   A scenario of galaxy formation in which large structures form first and then fragment to become galaxies.

Weak nuclear force   One of the four fundamental forces of nature. It controls the interaction of neutrinos.

WIMP   A weakly interacting massive particle. A generic term for a class of hypothetical particle which may form the missing mass. A form of non-baryonic cold dark matter.

World line   The trajectory of a body moving through spacetime.

Appendix 3

Bibliography

If you have read and enjoyed this book, I recommend any of those listed below, which have provided a great source of knowledge, inspiration and wonder. I have included comments about some of them, but the absence of a comment should not be interpreted as a rebuttal.

Arp, H., *Quasars, Redshifts and Controversies*, Interstellar Media, 1987. The personal account of one astronomer's work which has led him to believe that redshift is not solely a measurement of distance. Should be read by everyone before they subscribe to the Big Bang party line, as some of the evidence in here is hard to dismiss.

Barrow, J.D. and Tipler, F.J., *The Anthropic Cosmological Principle*, Oxford University Press, 1986. A provocative notion, well worth reading and guaranteed to stimulate discussion.

Berry, M., *Principles of Cosmology and Gravitation*, Adam Hilger, 1989.

Chincarini, G., Iovino, A., Maccacaro, T. and Maccagni, D., *eds., Observational Cosmology*, Astronomical Society of the Pacific Conference Series, 51, 1993.

Chown, M., *Afterglow of Creation*, Arrow, 1993. Tells the story of the scientific study of the microwave background from the point of view of the scientists involved. A good 'light' read.

Clark, S., *Redshift*, University of Hertfordshire, 1996. A thorough, non-technical examination of this phenomenon and what causes it, together with some of the problems in modern cosmology.

Coles, P. and Lucchin, F., *Cosmology*, Wiley, 1995. This is a post-graduate research tome. Very good, but not for the faint-hearted.

Evans, A., *The Dusty Universe*, Wiley–Praxis, 1995. Excellent account of baryonic dark matter within the Milky Way.

Foster, J. and Nightingale, J.D., *A Short Course in General Relativity*, Springer–Verlag, 1995.

Gribbin, J. and Rees, M., *The Stuff of the Universe*, Penguin, 1995. Another popular-level book which also introduces a little of the anthropic cosmological principle.

Hawking, S.W., *A Brief History of Time*, Bantam Press, 1988. The best-seller of popular cosmology books!

Illingworth, V., *ed.*, *Dictionary of Physics*, Penguin, 1990.

Kafatos, M. and Kondo, Y., *eds.*, *IAU Symposium 168: Examining the Big Bang and Diffuse Background Radiations*, Kluwer Academic Publishers, 1996.

Kitchin, C., *Journeys to the Ends of the Universe*, Adam Hilger, 1990. Well-written essays on some of the most difficult concepts in modern astronomy.

Lachiève-Rey, M., *Cosmology, a first course*, Cambridge University Press, 1995.

Leverington, D., *A History of Astronomy from 1890 to the Present*, Springer-Verlag, 1996. Excellent first point of reference for any historical perspective on astronomy.

Martin, J.L., *General Relativity, a first course for physicists*, Prentice Hall, 1996. A mathematical introduction to Einstein's work.

Mitton, J., *Dictionary of Astronomy*, Penguin, 1993.

Partridge, R.B., *3K: The Cosmic Microwave Background Radiation*, Cambridge University Press, 1995. Research-level thesis on every aspect of the cosmic microwave background. Very thorough for both observational work and theoretical understanding.

Pasachoff, J.M., Spinrad, H., Osmer, P.S. and Cheng, E., *The Farthest Things in the Universe*, Cambridge University Press, 1994. Four good introductory essays on aspects of cosmology.

Robson, I., *Active Galactic Nuclei*, Wiley–Praxis, 1996. Excellent undergraduate introduction to the field of active galaxies.

Roos, M., *Introduction to Cosmology*, Wiley, 1994. Perhaps a next step up from this book. Undergraduate level, but with a much greater proportion of mathematics.

Rowan-Robinson, M., *Cosmology*, Oxford University Press, 1996.

Rozental, I.F., *Big Bang, Big Bounce*, Springer-Verlag, 1988.

Silk, J., *The Big Bang*, W.H. Freeman, 1989.

Silk, J., *A Short History of the Universe*, Scientific American Library, 1994. Excellent book at a popular level, with just a hint of mathematics.

Tayler, R.J., *The Hidden Universe*, Wiley–Praxis, 1995. Tackles the whole problem of dark matter, effectively at undergraduate level.

Wagoner, R. and Goldsmith, D., *Cosmic Horizons: Understanding the Universe*, W.H. Freeman, 1982.

Wienberg, S., *The First Three Minutes*, Flamingo, 1996. Before *A Brief History of Time* was published, this was the classic.

Will, C.M., *Was Einstein Right?*, Oxford University Press, 1995. Excellent non-technical account of Einstein's theory of general relativity.

Wright, A. and Wright, H., *At the Edge of the Universe*, Ellis–Horwood, 1989.

Index